Royal Geographical Society

with the Institute of British Geographers

THE

GREAT

EXPLORERS

and their

JOURNEYS OF DISCOVERY

BEAU RIFFENBURGH

ANDRE
DEUTSCH

This book is dedicated with love to my parents, Ralph and Angelyn Riffenburgh, for their unflagging support, encouragement, and care over a period of many years.

ACKNOWLEDGEMENTS
Numerous people gave invaluable assistance in producing this book. I would especially like to thank my wife, Liz Cruwys, for all of her editorial input and continuing encouragement; Vanessa Daubney, for many, many different things, including pulling the entire project together, and making it actually happen; and Drew McGovern for his brilliant work on the sensational design of the book.

THIS IS AN ANDRE DEUTSCH BOOK

 The Royal Geographical Society
(with the Institute of British Geographers), London

Text © Wildebeest Publishing Ltd 2007

Design © André Deutsch Limited 2017

This edition published in 2017 by André Deutsch
A Division of the Carlton Publishing Group
20 Mortimer Street
London
W1T 3JW

The text and images in this book were first published in
The Royal Geographical Society Exploration Experience
(ISBN: 978 0 233 00219 4) in 2007.

The right of Beau Riffenburgh to be identified as the author of this work has been asserted in accordance with the Copyright, Designs and Patents Act 1988.

CONTENTS

EXPLORATION BEFORE THE GOLDEN AGE

THE AMERICAS

ASIA

AFRICA

AUSTRALIA AND THE PACIFIC

THE ARCTIC

ANTARCTICA

INTRODUCTION

Exploration is and always has been a fundamental part of human nature. All that "to explore" really means is to investigate systematically or to seek to ascertain something. It is thus closely related to basic human curiosity about an immensely broad range of topics. Physical scientists, those who study human or animal behaviour, medical researchers and cosmologists are as much explorers as those who in former times crossed mountains, deserts, seas, or realms of ice to uncover the geographical mysteries of the Earth.

However, geographical exploration and discovery produced a kind of personal and public excitement, thrill, and pride unmatched by even the most impressive and valuable scholarly advances in understanding either ourselves or any of the numerous features of the universe in which we exist. This is at least in part because so much of the success of geographical explorers or travellers was based on their individual strengths – body, will and character. They were driven forwards not simply by intellect, but by that incalculable quality perhaps best known as "heart". It is something that is not expressed to nearly the same degree in scientific exploration or even in space exploration, where man's personal bravery, physical determination, and heart are sublimated by intellectual and technological achievement.

However, even if different forms of exploration emphasized varying aspects of mankind's nature, they all were similarly driven: in part by curiosity, in part by the desire to show that people can make continual progress, and in part by the need to understand and even overcome the natural world. In fact, the concept of mankind's struggle against "Nature" and the ultimate "conquest of the world" – a persistent theme in numerous facets of life and an underlying force throughout the Western world in most of the nineteenth and early twentieth centuries – helped to create an obsession with filling in the white spaces on the map, which in turn led to this period becoming the "Golden Age" of exploration.

The individual explorers themselves were, of course, also driven in that Golden Age as they were throughout history by a variety of more personal goals. Fame and fortune were long the hoped-for rewards not only of the men and women who journeyed to new places but those who backed their enterprises. Whether it was the Spanish Conquistadors of the fifteenth and sixteenth centuries, the hardy men searching for new whaling and sealing grounds, or the likes of the reporter Henry Morton Stanley seeking David Livingstone at the order of the owner of the largest newspaper in the United States, the goal of riches frequently lurked in the background. At various points, exploration was also the product of religious proselytizing, national pride, a desire to escape from a rapidly industrializing world to a more "pure" environment, or a necessary precursor to the development of empire. And scientific study became a progressively more important aspect, perhaps particularly in the Antarctic, where it became and remains the major focus of human involvement.

Today, there is little left to explore geographically on Earth, and it has been argued that the world's last true, classic explorer was Wilfred Thesiger, who died in 2003. Nevertheless, the concept of exploration still holds great appeal, and many adventurers only too willingly present themselves as explorers. In one way, perhaps, they are – in that they continue to explore the limits of human endurance. This book offers an introduction to the experience of true geographical exploration, and brings the reader even closer to those who risked their lives investigating unknown areas by its inclusion of facsimiles of documents (many of which come from the Royal Geographical Society's archives) that were produced by them and testify to their intrepidity.

Beau Riffenburgh

EXPLORATION BEFORE THE GOLDEN AGE

EXPLORATION BEFORE COLUMBUS

In one sense, exploration is as old as mankind: early peoples spread out from Africa and ultimately populated most of the world. For ancient civilizations, exploration was usually a by-product of military conquest or the establishment of trading routes. One of the earliest known explorers was an Egyptian, Harkhuf, who lived during the Sixth Dynasty of the Old Kingdom. Around 2270 BC, he led several expeditions up the Nile; on the last one, he brought back a "dwarf" to Pharaoh Pepi II. Some 800 years later, the Egyptian queen Hatshepsut sent a fleet down the Red Sea to the land of "Punt", although it is uncertain where Punt actually was.

OPPOSITE The Royal Geographical Society commissioned this copy of the original Hereford "Mappa Mundi" (a mediaeval European world map) in 1830–31. The thirteenth-century original, on display in Hereford Cathedral, England, is drawn on vellum and measures 158cm x 153cm. Jerusalem is shown at the centre of the circle, east is on top, with the Garden of Eden in a circle at the edge of the world. Europe is in the bottom left – Scotland, England and Ireland are drawn as large islands at the north-western border – while Africa is on the right.

From a European perspective, much of the world was opened up by Phoenician traders in the Mediterranean or by the conquests of Alexander the Great in Asia. But exploration was not all initiated in Europe – the Chinese had a long history of such expeditions. From around 138 BC to 116 BC, Zhang Qian travelled as far as Samarkand, but his journeys were interrupted for a decade when he was detained by the Xiongnu. At the beginning of the fifth century AD, the Buddhist traveller Faxian became the first Chinese known to reach India, following the Silk Road across the Pamirs and then turning south, eventually arriving in Sri Lanka. Two hundred years later, Xuanzang made a similar, but more extensive journey, following which he wrote one of the most important early Chinese travel accounts.

Exploration also flourished in the Islamic world, which extended at its height from Spain and West Africa to central Asia and beyond. One of the first notable Arab explorers was Suleiman the Merchant, who in 850–51 sailed from the Persian Gulf to India, the Spice Islands, Vietnam and on to China; his later accounts had a profound effect on many Arabic and Persian geographers. The next century Ibn Fadlan reached the kingdom of the Bulgars on the Volga River.

From the far north, the Vikings, or Norsemen, spread far and wide from the eighth century, settling in Normandy, founding a state at Kiev, and conquering Sicily. They also discovered a host of new lands to the west, settling the Shetlands and the Faeroes by the early ninth century, and Iceland shortly thereafter. In 982, Erik Thorvaldsson ("Eric the Red") set out from Iceland to seek a land reported approximately a century earlier by Gunnbjörn Ulfsson. Eirik "rediscovered" it, named it Greenland, and started its colonization. The Vinland sagas

LEFT The Polos *en route* to China, a scene from a fourteenth-century map prepared for Charles V of Spain by a Catalan cartographer.

IBN BATTUTA

Perhaps the greatest traveller ever, Ibn Battuta was born in Tangier in 1304, and at 21 joined a pilgrimage for Mecca. It was the first of eight expeditions lasting 29 years and covering 120,000 km (75,000 miles). His initial journey included a detour down the Red Sea and visits to Damascus, Mecca and Baghdad. A subsequent voyage took him around Arabia and south to Somalia and Tanzania. His greatest expedition, lasting 16 years, included Constantinople, southern Russia, the fabled cities of Bukhara and Samarkand, and Afghanistan, India, the Maldives and China. Heading back to Tangier, he made diversions to Granada and south through the Sahara to the fabled city of Timbuktu.

RIGHT A scene from the Hajj, the vast pilgrimage to Mecca that is required of every Muslim at least once in his life.

differ regarding who was the first European to reach North America. According to *Groenlendinga saga*, around 986 Bjarni Herjólfsson sighted coastlines that were probably those of Newfoundland, Labrador and Baffin Island. But according to *Eiríks saga*, that honour belonged to Leif Eiriksson, who in 1001–02 reached areas he named Helluland, Markland and Vinland. A settlement was soon established at the last of these, and although temporary, it was the first European habitation in North America.

Perhaps the greatest European traveller before Columbus came from a family of Venetian merchants. Marco Polo's father, Nicolò, and uncle, Maffeo, had engaged in remarkable commercial travels, visiting the mysterious khanates of Bukhara and Samarkand, and continuing to the court of Kublai Khan in the dazzling city of Beijing. Their

journey had taken more than six years – perhaps even ten – but in 1271, not long after their return, they set out again, taking along 17-year-old Marco.

Through the lands of the Turks, Armenians, Persians and Afghans they travelled, before crossing the Hindu Kush and following the Silk Road to China. There, the Polos served the Great Khan for about 17 years, during which Marco visited Mongolia, Burma, India and Malaysia. They finally departed for Europe, escorting a Mongol princess to Persia via Sumatra and the Indian Ocean, before reaching Venice in 1295. Several years later, Marco Polo was captured in a conflict with Genoa and imprisoned. One of his fellow prisoners was a writer named Rustichello, who convinced Marco to dictate the story of his travels, which, when published, opened up to Europe the many wonders of what the Polos called Cathay.

The Caravan.

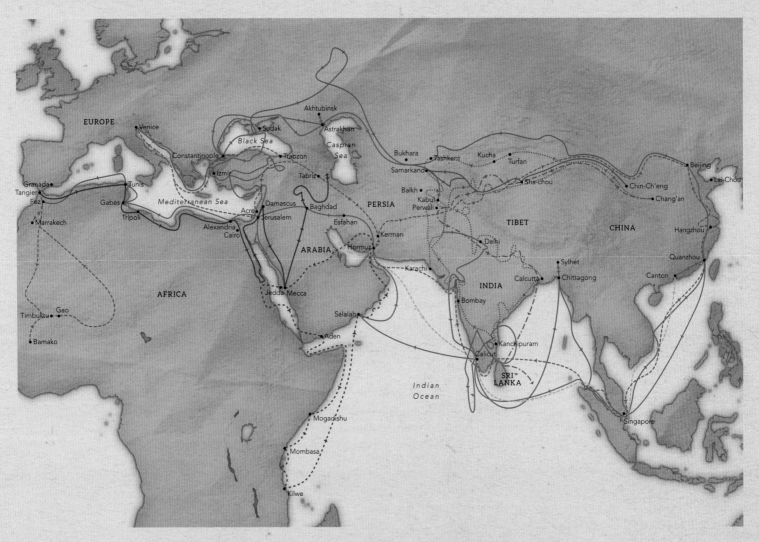

Zhang Qian, 128–116 BC

Faxian, AD 399–414

Xuanzang, 629–645

Suleiman the Merchant, 850–851

Maffeo and Nicolò Polo
outward route, 1260–1266

Marco Polo, 1271–1295

Ibn Battuta, 1325–1327

Ibn Battuta, 1328–1330

Ibn Battuta, 1330–1346

Ibn Battuta, Uncertain route

Ibn Battuta, 1346–1349

Ibn Battuta, 1349–1354

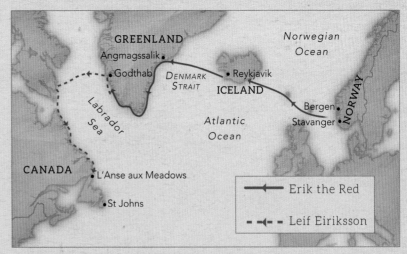

Erik the Red

Leif Eiriksson

WHERE WAS VINLAND?

When Leif Eriksson reached North America, he wintered in a place he named Vinland, before returning to Greenland bearing with him grapes and vines. For many years, the exact location of Vinland was a matter of great dispute. Then, in 1961, the Norwegian scholars Helge and Anne-Stine Ingstad began excavations at L'Anse aux Meadows in northern Newfoundland. The foundations of eight buildings were discovered, including a large house almost identical to Leif's "great hall" in Greenland. It is now generally accepted that this was indeed Leifsbudir, Leif's settlement in Vinland.

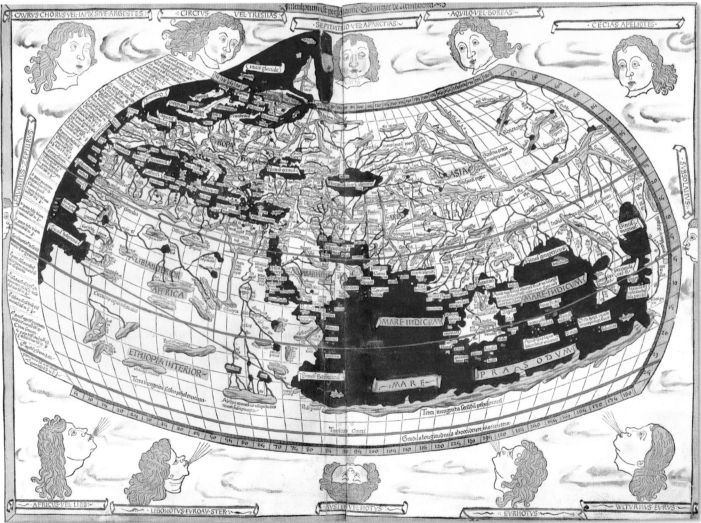

TOP A statue of Leif Eriksson that stands before the Hallgrimskirkja in Reykjavik. It was presented to Iceland by the US Congress in 1930.

ABOVE The map of the world produced by Greek geographer and astronomer Ptolemy in Alexandria around 140 AD. At the time his eight-volume *Guide to Geography* was the greatest compilation of geographical information collected, and this map influenced travellers for centuries.

CHRISTOPHER COLUMBUS

No expedition in history is more famous than the one that culminated on 12 October 1492, when three Spanish ships arrived at a tiny island in the Bahamas. None of the crew then realized that they had discovered a part of the world previously totally unknown to Europeans, but it was, nevertheless, a moment of triumph for the commander of the small flotilla, Christopher Columbus.

TOP Columbus' flagship *Santa Maria* (centre) accompanied by the two caravels, *Pinta* and *Niña*. The size of *Santa Maria* has been estimated at approximately 100 tons, with each of the two smaller ships only half that.

OPPOSITE Columbus is received by Ferdinand and Isabella of Spain in 1493, on return from his first voyage. He presented several New World natives to the monarchs, as well as a variety of plants and foods.

Born in Genoa in 1451, the son of a weaver, Columbus (known as Cristoforo Colombo in Italy and Cristóbal Colón in Spain) went to sea at 14. Myths abound about his early years as a mariner, but what is certain is that in the 1470s he moved to Lisbon – then a centre of maritime commerce – and that he also lived for a period in Madeira. As he gained practical experience in the Atlantic, Columbus became convinced that it was possible to reach the Far East by sailing westwards, and in 1484 he proposed a plan for doing so to King John II of Portugal, who rejected it. Undaunted, Columbus turned to the Spanish monarchs Ferdinand and Isabella, and began a six-

year campaign to gain their patronage, in which he was ultimately successful.

On Friday 3 August 1492, Columbus set sail from the Spanish port of Palos in command of three ships, the flagship *Santa Maria* and two small caravels, *Pinta* and *Niña*, captained by the brothers Martin and Vicente Pinzón respectively. They continued on 6 September after restocking at the Canary Islands, but following 13 days of strong winds they slowed to a pace that made the crews exceedingly nervous. On 10 October Columbus promised to turn around in three days if no land was sighted, but two days later the ships arrived at the island that Columbus named San

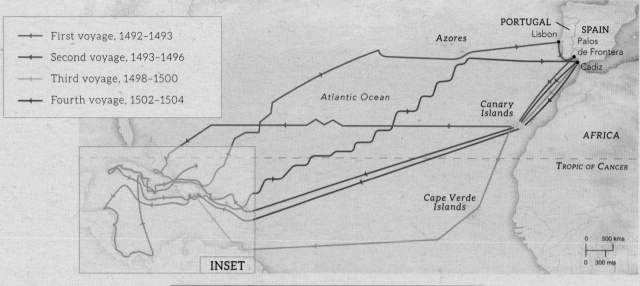

- ➤ First voyage, 1492–1493
- ➤ Second voyage, 1493–1496
- ➤ Third voyage, 1498–1500
- ➤ Fourth voyage, 1502–1504

PORTUGAL
Lisbon
SPAIN
Palos de Frontera
Cadiz
Azores
Atlantic Ocean
Canary Islands
AFRICA
TROPIC OF CANCER
Cape Verde Islands
INSET

0   500 kms
0   300 mls

San Salvador Island
CUBA
Navidad
Isabela
HISPANIOLA
PUERTO RICO
JAMAICA
Santo Domingo
DOMINICA
MARTINIQUE
TRINIDAD
Portobello
Belen
VENEZUELA

Salvador. He took possession in the name of Spain, and made his way south to Cuba, which he assumed was part of mainland Asia. Columbus then discovered the island of Hispaniola, but on Christmas Eve *Santa Maria* was wrecked, so he left a party ashore and sailed in his two smaller ships back to Spain, arriving on 15 March 1493, to a hero's welcome.

Columbus's second voyage (1493–96) was more ambitious, with 17 ships – three carracks and 14 caravels – sailing from Cadiz with more than 1,000 men. After some five weeks they sighted the island of Dominica, and then discovered Guadeloupe and Puerto Rico, before proceeding to Hispaniola to found the coastal settlement of Isabella. Columbus explored the south coast of Cuba and discovered Jamaica before a series of quarrels with his subordinates prompted him to head back to Spain.

Despite what had proved an obvious lack of ability as a political administrator, in 1498 Columbus was sent on a third voyage, as the Spanish monarchs were

NAVIGATIONAL INSTRUMENTS

During Columbus's time there were two common navigational tools to help fix positions when ships were out of sight of land: the quadrant and astrolabe. The more primitive was the quadrant, although refined versions remained in use for centuries. It was triangular, with a curved bottom side and a lead weight hanging on a string from the top angle. The astrolabe was circular with degrees marked on its edge and a movable sighting bar. Both were used to determine the height of the Sun or stars, measurements that, when compared to readings taken in port, could be used to calculate latitude.

ABOVE An astrolabe from the holdings of the Royal Geographical Society. The centre part revolved to allow bearings of the stars or planets to be taken.

concerned about other countries eyeing territories in the New World. Three ships – which included 30 women for the settlers from the previous expedition – were sent directly to Hispaniola, while Columbus, with three more ships, followed a more southerly route than previously. In late July, he discovered the island of Trinidad and then reached the South American mainland along the coast of Venezuela. But he continued to have problems with his subordinates, and in 1500 a new royal governor shipped him back to Spain in chains.

Still in favour with Queen Isabella, the explorer was allowed to return to the New World on a fourth voyage in 1502, but was forbidden to approach Hispaniola. Looking for a passage that would take him to India proper, he sailed to the west of Cuba and along the coasts of what are today Honduras, Nicaragua, Costa Rica and Panama. Returning east, he was shipwrecked on Jamaica, where he was marooned for a year before being rescued. Columbus then returned to Spain in November 1504, where he lived unhappily for a year and a half before dying in May 1506.

RIGHT An extract from the letter written by Christopher Columbus to Ferdinand and Isabella upon returning from his first voyage to the New World. The letter summarized his journal of the voyage, and was widely printed in both Spanish and Latin. (See Translations, page 204.)

THE FIRST SETTLEMENTS

On Christmas Day 1492, the day after *Santa Maria* ran aground at Hispaniola, Columbus interpreted the mishap as a sign to found a colony there. He therefore left behind about 40 men, equipped with stores and ammunition. He named the place Navidad (the Spanish for Christmas). The next year, Columbus discovered that the settlers had all been massacred, but in 1494 he founded another settlement 50 kilometres (30 miles) farther east, named Isabella. This, too, suffered immediately, both from hostilities with the natives and Columbus's heavy-handed rule, but it became the first European town in the New World.

LEFT A woodcut from a 1493 edition of Columbus' letter. Shown in unrealistically glorious fashion is the fort that he established on Hispaniola. On his return the next year, Columbus discovered the colony had been wiped out.

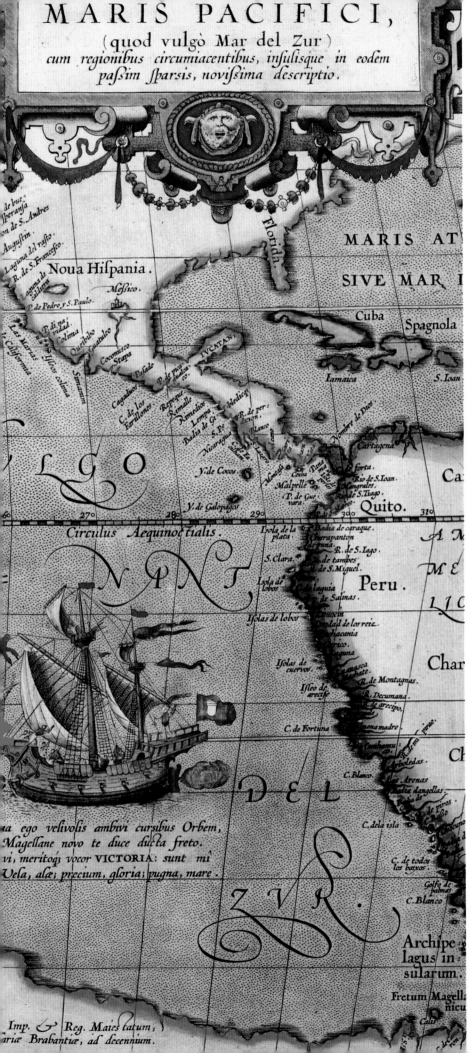

On the map:

MARIS PACIFICI,
(quod vulgò Mar del Zur)
cum regionibus circumiacentibus, insulisque in eodem
passim sparsis, novissima descriptio.

Noua Hispania.

Messico.

MARIS AT

SIVE MAR I

Cuba

Spagnola

Iamaica

S. Ioan

Quito.

Circulus Aequinoctialis.

Peru.

Char

ia ego velivolis ambivi cursibus Orbem,
Magellane novo te duce ducta freto.
vi, meritoq3 vocor VICTORIA: sunt mi
Vela, alæ; precium, gloria; pugna, mare.

Archipe
lagus in
sularum.

Fretum Magella

Imp. & Reg. Maiestatum,
ariæ Brabantiæ, ad decennium.

THE GREAT VOYAGERS

Throughout most of the fifteenth century, it was one of Europe's smaller kingdoms, Portugal, that led a great wave of maritime exploration. The key figure behind this push was Prince Henry "the Navigator", the third son of John I of Portugal. Shortly after the capture of the Moorish city of Ceuta in North Africa in 1415, Henry set up a school for navigation and began promoting the patient, systematic exploration of the coast of West Africa. This process led eventually to Portuguese dominance of the Atlantic slave trade (which developed from the 1440s) as well as to the discovery of a seaborne route to India.

17

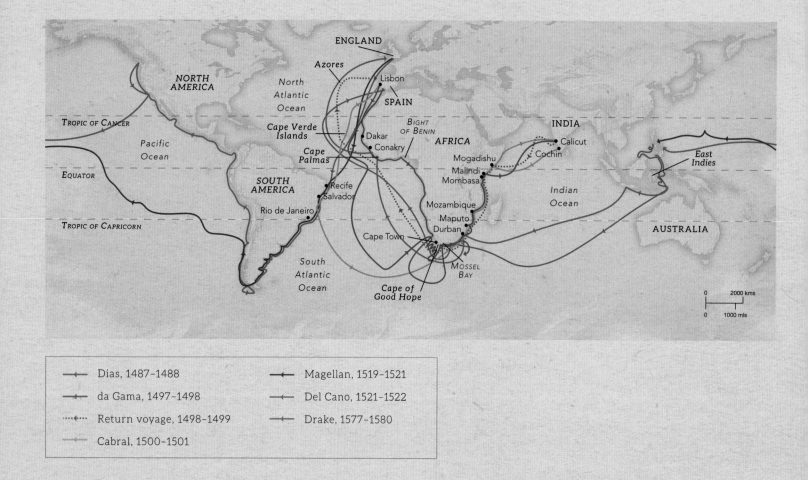

→ Dias, 1487–1488	→ Magellan, 1519–1521
→ da Gama, 1497–1498	→ Del Cano, 1521–1522
···▶ Return voyage, 1498–1499	→ Drake, 1577–1580
→ Cabral, 1500–1501	

Henry died in 1460, but the investigation of the African coast continued under Fernão Gómes, a Lisbon merchant, who agreed to explore at least 100 leagues a year in return for a trading monopoly. In the 1480s, expeditions sent by King John II made even more rapid progress. In 1482, Diego Cão reached the Congo River. Then, in 1488, Bartolomeu Dias, having headed directly south from what is now Namibia, reached a point south of the continent before turning northeast and landing at Mossel Bay. Having thereafter established that the coast turned northeast, he was forced by a discontented crew to return home, discovering Cape Aghulas at the southern tip of Africa and the Cape of Good Hope on the way.

In 1497–98, a sea route to the East was finally established. Rather than hugging the African coast and struggling through poor winds in the Gulf of Guinea as his predecessors had, Vasco da Gama sailed south from the Cape Verde Islands, and turned east when at the latitude of southern Africa. He then sailed north up the east African coast, while attempting to gain relief for his scurvy-ridden crew, which had been out of the sight of land longer than any previous expedition. At what is today the coast of Kenya, da Gama turned east to reach India, where he obtained pepper, cinnamon and cloves, which he took back to Portugal, arriving in July 1499.

Da Gama's success inspired further expeditions which followed the mid-Atlantic route. On one, the nobleman Pedro Alvares Cabral, on his way to India, was blown far to the southwest, landing in Brazil in April 1500. He claimed it for Portugal, and then proceeded to India. Despite the unexpected detour, it was the fastest journey from Europe to India prior to the invention of the steamship.

Another Portuguese nobleman who sailed to India via Africa – and on to the Moluccas or Spice Islands – was Ferdinand Magellan. He believed he could reach the Moluccas via a passage through the Americas and

OPPOSITE TOP This is a nineteenth-century copy of a map drawn in 1500 by Juan de la Costa, a Spanish cartographer, explorer and conquistador, who sailed with Christopher Columbus on his first three voyages to the New World. The map has coloured insets of significant rulers and buildings, and an impressively detailed list of coastal place names, particularly on the west coast.

OPPOSITE LEFT An early drawing of the vicinity of Table Bay and the Cape of Good Hope. Although the Portuguese were the first Europeans to reach the region, it was the Dutch East India Company that first established a station there in 1652.

OPPOSITE RIGHT Vasco da Gama listens to the entreaties of his frightened and scurvy-ridden crew to return to Portugal. Da Gama refused to turn back, and continued up the east coast of Africa and thence on to India.

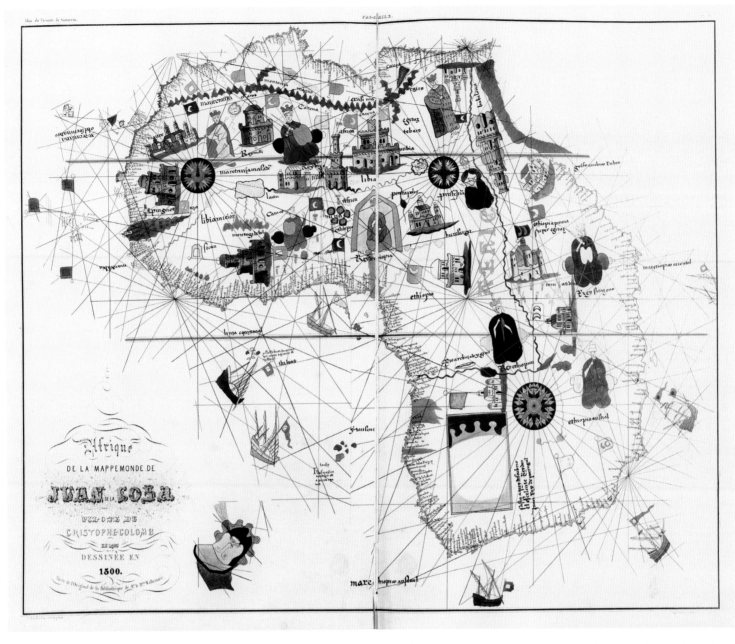

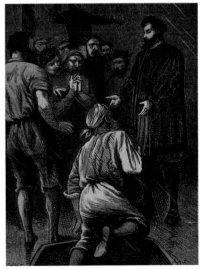

thence across the Pacific Ocean, which Vasco Núñez de Balboa had discovered in 1513. Having lost favour in Portugal, Magellan offered his services to Charles V of Spain.

Magellan left Spain with five ships in September 1519, and sailed down the coast of South America, investigating its major rivers *en route*. After wintering at San Julian (in Patagonia) and overcoming a mutiny, the wrecking of one ship and the desertion of another, he made his way through the Strait of Magellan and entered the Pacific in November 1520. In the next three months, the remaining ships crossed the Pacific, many men dying of scurvy before they reached Guam. Magellan was killed in the Philippines in April

LEFT AND BELOW Ferdinand Magellan, a Portuguese navigator in the service of Spain. Magellan made the mistake of becoming involved in a local war in the Philippines, and was killed attempting to seize the island of Mactan.

1521, but two ships continued under Juan del Cano.
In September 1522, the last ship, *Vittoria*, reached
Spain with just 18 of the armada's original 250
men remaining, having completed the first voyage
around the world.

The second circumnavigation took place half a
century later under the Englishman Francis Drake.
Intent on raiding the Spanish in Peru, Drake left
Plymouth in 1577 and followed the South American
coast, passing through the Strait of Magellan. Once
in the Pacific, he was blown far south by a storm,
discovering the Drake Passage and proving that
Tierra del Fuego was an island, not part of a great
southern continent. Drake then sailed up the west
coast of the Americas, reaching San Francisco
before setting off across the Pacific. With a cargo of
cloves from the Spice Islands and plundered Spanish
treasure, he continued west and reached England in
September 1580.

MEASURING SPEED AT SEA

If mariners could estimate their speed at sea, they could determine their actual position with greater accuracy. One common way to measure a ship's speed was to tie a rope marked by a series of knots at regular intervals to a log, which was then thrown overboard. The speed with which the knots went over the side of the ship gave an estimation of how fast the ship was travelling. The term "knots", used today to measure a ship's speed, came from this practice.

ABOVE The men aboard a nineteenth-century ship go through the laborious process of finding out at how many knots they are travelling.

ABOVE Henry the Navigator portrayed around 1450. Along with King John II, he was largely responsible for the dominance of Portuguese maritime exploration.

URDANETA'S PASSAGE

After Magellan's circumnavigation and the establishment of Spanish colonies on the New World's west coast, numerous attempts were made to cross the Pacific and then return to the Americas. However, these were thwarted by ill luck and northeast trade winds. Finally, in 1565, Andrés de Urdaneta sailed north from the Philippines and found the northeasterly Japan current, which led to the easterly North Pacific current. These helped him to reach California and then follow the coast south to Mexico. What became known as Urdaneta's Passage was used for two centuries.

RIGHT The tiny ships of the Spanish, Portuguese, Dutch and English explorers would not receive Board of Trade approval today.

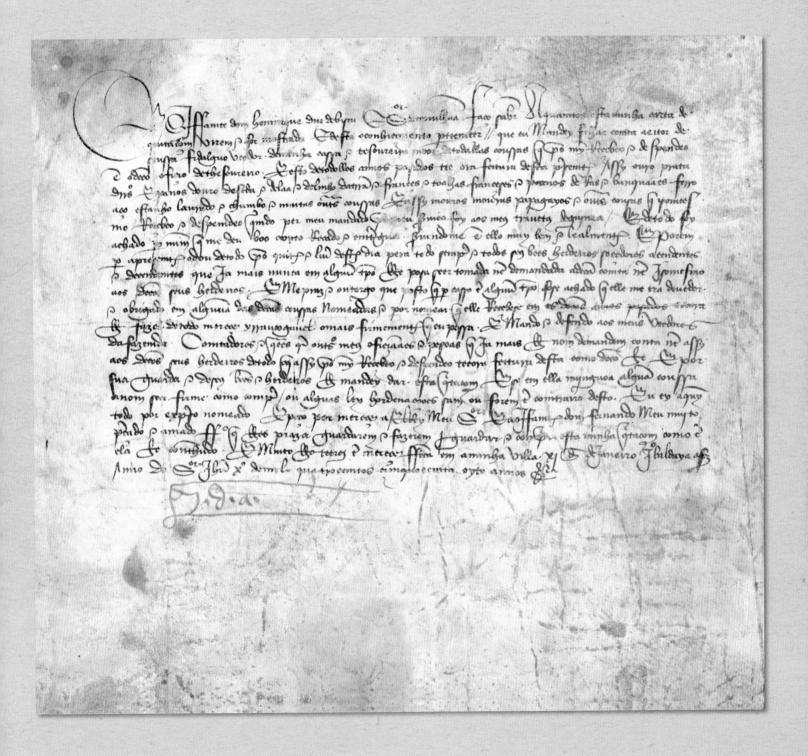

ABOVE A letter drawn up for Henry the Navigator two years before he died. The Prince's munificence with the possessions gained on the voyages he sponsored was typical. Once, when men from his household died on an expedition, he gave generous pensions to their wives and children. (See Translations, page 204.)

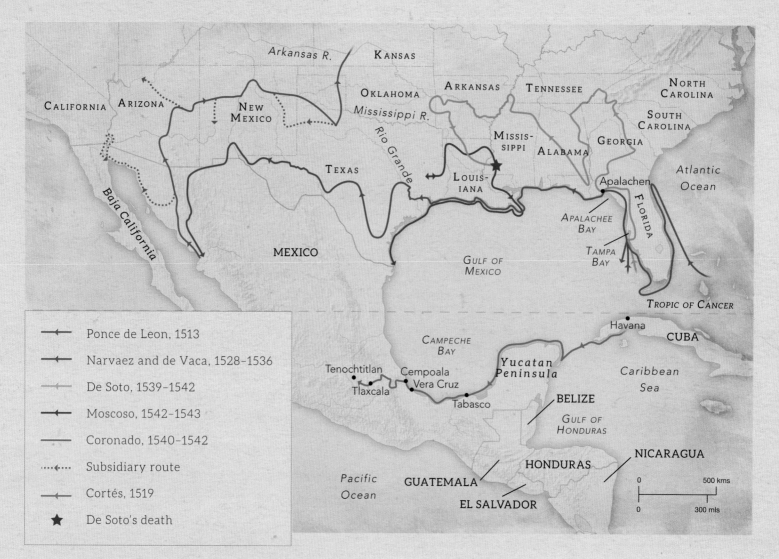

Arkansas R. KANSAS

CALIFORNIA ARIZONA NEW MEXICO OKLAHOMA Mississippi R. ARKANSAS TENNESSEE NORTH CAROLINA

SOUTH CAROLINA

Rio Grande MISSIS-SIPPI ALABAMA GEORGIA

TEXAS Atlantic Ocean

Baja California LOUIS-IANA Apalachen FLORIDA

MEXICO GULF OF MEXICO APALACHEE BAY

TAMPA BAY

TROPIC OF CANCER

Havana CUBA

CAMPECHE BAY

Tenochtitlan Cempoala Yucatan Peninsula Caribbean Sea

Tlaxcala Vera Cruz

Tabasco BELIZE Gulf of Honduras

HONDURAS NICARAGUA

Pacific Ocean GUATEMALA

EL SALVADOR

⟵	Ponce de Leon, 1513
⟵	Narvaez and de Vaca, 1528–1536
⟵	De Soto, 1539–1542
⟵	Moscoso, 1542–1543
—	Coronado, 1540–1542
⋯⟵	Subsidiary route
⟵	Cortés, 1519
★	De Soto's death

0 ——— 500 kms
0 ——— 300 mls

THE SPANISH IN NORTH AMERICA

Despite the glory and riches that voyages of Columbus seemed to promise, two decades later the Spanish had still not investigated the mainland of North America. The first stage in that exploration only began in 1513, when Juan Ponce de León left Puerto Rico to locate the reputedly rich island of Bimini, which was also reported to be the home of the legendary "Fountain of Youth". In April, he discovered what he thought to be an island, which he named Florida for the Spanish Feast of the Flowers at Eastertide, Pascua Florida. He returned to Puerto Rico to obtain a royal licence to settle and govern this new land. However, when de León returned to Florida in 1521, he was fatally wounded in an attack by Indians.

OPPOSITE Francisco de Coronado's massive expedition never found the legendary Seven Cities of Cibola, but it did introduce horses to the native peoples of the American west.

TOP Montezuma I attends a human sacrifice in Tenochtitlan. The Spanish were horrified at the seeming insatiability of the Aztec gods for blood. An early Spanish chronicler recorded that more than 50,000 people a year were sacrificed.

Meanwhile, in 1519, Hernán Cortés, who had served under Diego de Velásquez in the conquest of Cuba, was sent to find the rich kingdom said to lie to the west of the recently discovered Yucatan Peninsula. With some 500 men, 16 horses and six cannon, he landed on the Mexican coast and founded the city of Vera Cruz. In August, he began making his way from the tropical Gulf Coast through the high mountains and plains towards the great city of Tenochtitlan – located on an island in a lake – from where Montezuma ruled the Aztec empire. Along the way, Cortés persuaded peoples previously conquered by the Aztecs to join him.

The Spaniards entered Tenochtitlan and seized Montezuma, making him a Spanish puppet. This situation lasted five months, until Cortés temporarily left the city, and the Aztecs rose against the men he left behind. Montezuma was killed by his people while trying to maintain peace. Cortés suffered an initial defeat, but he reorganized his forces and the Spanish laid siege to Tenochtitlan, sacked it, and executed the last Aztec ruler. Cortés's New Spain became the centre of future exploration to the north, south and into the Pacific.

In 1528, Pánfilo de Narváez, who had participated in the invasion of Cuba, led an expedition to explore Florida, of which he had been appointed governor. He landed in the south of the peninsula, and heard tales of a city of riches, but a march to the north uncovered no such wealth. Retreating to the Gulf Coast, his men built five boats and headed west until a storm smashed the vessels, throwing the survivors on to an island near Galveston, Texas. With Narváez lost at sea, a junior officer, Álvar Núñez Cabeza de Vaca, and several other men began a terrible and remarkable journey west. Twice, they were enslaved by Indians for more than a year, although they were given aid by other tribes. They slowly made their way west across Texas and northern Mexico close to the Gulf of California, before reaching Mexico City with the help of Spanish slave traders in 1536, eight years after their departure.

Cabeza de Vaca's written accounts helped inspire Hernando de Soto, a bloodthirsty conquistador who had been in Peru with Francisco Pizarro (see pp 35–37), and whose expedition in 1539–43 terrorized the native population throughout what became the southeastern United States. Starting with 600 men, he marched through Florida, Georgia, and the Carolinas before turning west, crossing the Appalachian Mountains, and meandering through Tennessee, Alabama, Mississippi and Louisiana, all the while killing or enslaving local people. In 1541, his men became the first Europeans to see the Mississippi River, which they crossed before trekking through Arkansas and Kansas. In 1542, De Soto died, and his party, under Luis de Moscoso, returned to Mexico.

At the same time, in 1540–42, Francisco de Coronado traversed areas farther west while searching first for the fabled Seven Cities of Cibola, and then for the legendary Quivira. Coronado divided his expedition into sections, which allowed his men to reach Kansas, Colorado, the Painted Desert in New Mexico and the Grand Canyon in Arizona. However, nowhere amongst the Pueblo peoples of the American west did Coronado find any riches. He brought back only geographical knowledge, which was considered of little value.

THE CONQUEST OF THE CARIBBEAN

The Spanish took little interest in Caribbean islands other than Hispaniola in the years immediately after Columbus reached the New World. But in 1508, Juan Ponce de León landed on Puerto Rico to subjugate the native peoples and explore and colonize the island, of which he then became governor. Three years later, Diego de Velásquez, who had sailed on Columbus's second voyage, led an expedition from Hispaniola to conquer Cuba. He landed at its east end, established the first permanent European settlement, Baracoa, and, within three years, crushed local resistance to Spanish rule.

LEFT Diego de Velásquez's brutal conquest of Cuba was far from the light-hearted scene shown in this romanticised painting.

THE CONQUISTADORS

The Conquistadors – a name which comes from the Spanish word for conqueror – were rugged adventurers and soldiers of fortune who came to the New World to gain gold and fame and, for some, to serve the king of Spain or the Catholic Church. Few in overall number, the Conquistadors tended to demonstrate a fanatical drive and an insatiable lust for the riches they plundered from the native peoples, frequently with little mercy and great brutality. The most famous Conquistadors were Hernán Cortés and Francisco Pizarro.

RIGHT Cortés triumphantly enters Tlaxcala, the home of his allies, after defeating the Aztecs at Otumba in July 1520.

BELOW The plan of Tenochtitlan, capital of the Aztecs, as attributed to Hernán Cortés. The map shows the city's place in the centre of the lake and the causeways that allowed the Spaniards both to enter the city and to escape when necessary.

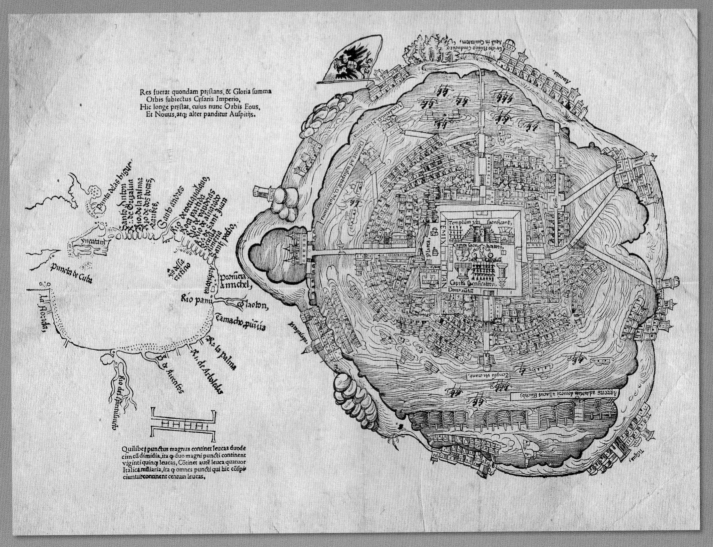

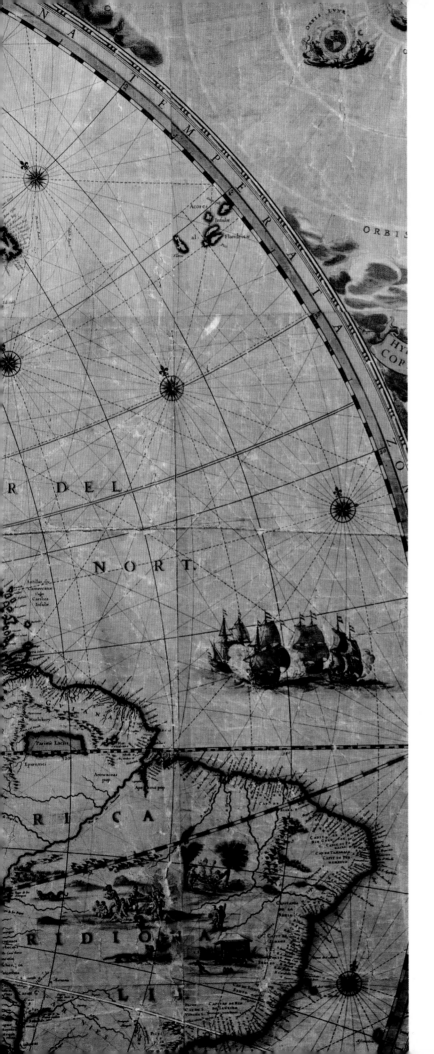

LEFT The Nova Totius Terraum Orbis Tabula was a world map drawn in two hemispheres by Joan Bleaeu, the Dutch cartographer who also produced the many-volumed Atlas Major. This map shows the Americas and, most notably, California as an island. This cartographic error, which persisted for about 100 years after 1622, began when Spanish reports claimed that a great bay in the north connected to the mouth of the Baja River. Interestingly, the Atlas Major includes a world map showing California as an island, but also an earlier map of America, first engraved in around 1619, correctly as a peninsula.

THE FRENCH IN NORTH AMERICA

While Spanish conquistadors lusted for gold, other Europeans were drawn to more northerly parts of the New World by their desire to find a sea route to China and take advantage of the remarkable commercial opportunities such a route would open up. After his voyage of 1497, John Cabot reported the existence of vast stocks of cod off Newfoundland, a discovery which precipitated a surge of voyages to exploit the massive fishery, during which, in turn, new discoveries were made but rarely recorded.

For years, the investigation of the Atlantic seaboard was carried out primarily by Portuguese, Breton and Norman fishermen; serious explorers only entered the scene in 1524, when a French expedition under the Florentine navigator Giovanni da Verrazano attempted to find a route to the Pacific. Verrazano made landfall near Cape Fear, North Carolina, and then sailed north to Newfoundland, investigating nooks and crannies along the way, but leaving important inlets such as Chesapeake Bay and the Hudson River unexplored. He made two further voyages; on the final one, Verrazano landed by himself on the island of Guadeloupe, where his horror-struck crew watched Carib Indians kill him, carve him up and eat him.

Verrazano's failure to discover a western passage led later French expeditions to concentrate on higher latitudes. In 1534, Jacques Cartier explored the Newfoundland coast and the Gulf of St Lawrence. The next year, he penetrated 1,600 kilometres (1,000 miles) up the St Lawrence River, discovering the Iroquois Indian village of Hochelaga – where Montreal would be founded – before being stopped by violent rapids. His party wintered at the St Charles River, but many of his men died from cold and scurvy before being shown by the Iroquois that the bark and leaves of the white cedar could produce a drink that cured the dreaded sickness. In 1541, Cartier made a third trip, but progressed little farther than before.

It was another 60 years before Samuel de Champlain significantly extended Cartier's discoveries. Champlain led a series of expeditions in the opening decades of the seventeenth century that explored and mapped the Atlantic coast south to Massachusetts and inland from the St Lawrence, both to the north and south. But the man known as

ABOVE Champlain firing his arquebus at the Iroquois in 1609. Champlain championed the Huron against their ancient enemies, leading to over a century of hostility between French and Iroquois.

OPPOSITE LEFT Letter Patent from François I for Jacques Cartier – the "Captain General" – to colonize the region near modern-day Montreal. Before the expedition left in 1541, Cartier was replaced as leader. He ignored the new commander, but brought back very little wealth or geographical information. (See Translations, page 204.)

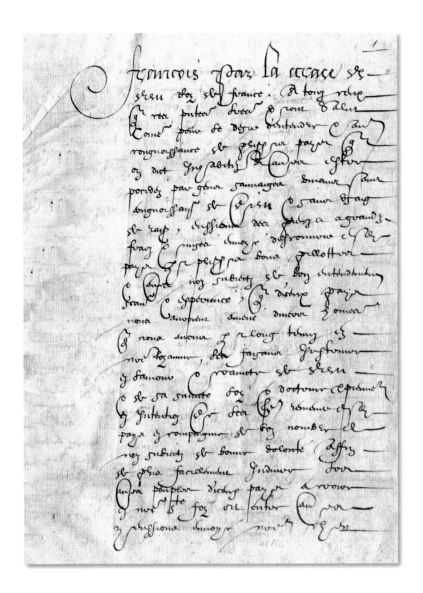

the "Father of New France" did more than explore: he established an outpost that became Québec City; he returned to France to encourage interest in New World expansion; he fascinated the public with his accounts; and he served as mentor for other young French explorers. These included Etienne Brûlé, who mapped much of the Great Lakes region, and Jean Nicolet, the first European extensively to explore Lake Superior.

Exploration south of the Great Lakes continued in 1672, when fur-trader Louis Jolliet was commissioned to investigate Indian stories of a great river. Among his party was the Jesuit priest and interpreter Jacques Marquette. Departing from the western shore of Lake Michigan, they traced a system of rivers and lakes into the upper reaches of the Mississippi. The pair continued south past the confluences of the Missouri and Ohio rivers, encountering previously unknown

RIGHT TOP Samuel de Champlain, who spent 15 years investigating little-known areas in the New World, was far more than just an explorer. His work as a colonizer – which continued over three decades – laid the foundation of France's North American empire.

RIGHT MIDDLE Jacques Cartier, who had been commissioned by François I to lead an expedition, initially established good relations with Indians, but damaged them by erecting a 30-foot cross that claimed the region for France.

RIGHT BOTTOM A glazed terra-cotta bust of Giovanni da Verrazano. Sailing from Le Havre, he reached the area of New York City in April 1524. In the 1960s the Verrazano Narrows Bridge was named in his honour.

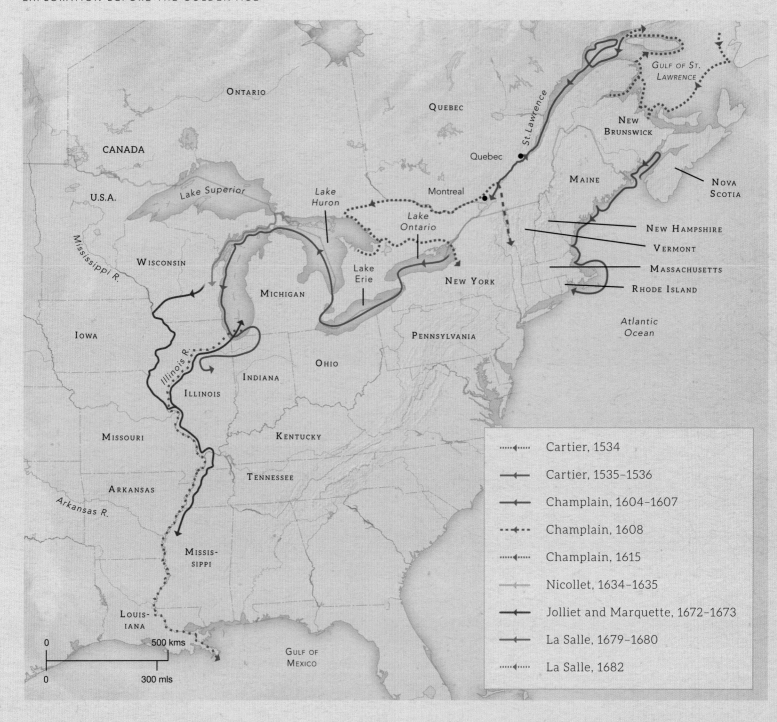

Cartier, 1534

Cartier, 1535–1536

Champlain, 1604–1607

Champlain, 1608

Champlain, 1615

Nicollet, 1634–1635

Jolliet and Marquette, 1672–1673

La Salle, 1679–1680

La Salle, 1682

Indian tribes, wildlife and flora. Finally, near the Arkansas River, just north of Spanish territory, they turned back, having been assured that the river continued to the Gulf of Mexico.

However, it took another decade for René-Robert Cavelier, Sieur de La Salle, to prove the destination of North America's greatest river. In 1679, he sailed through Lakes Erie, Huron, and Michigan, and thence into the waterways south of Lake Michigan, reaching the confluence of the Illinois River and the Mississippi. After returning to Montreal, in 1681

La Salle went south again, continuing downriver past Jolliet's turning point and, on 6 April 1682, reaching the Gulf of Mexico. He then claimed the river's entire catchment area for France, naming it Louisiana, and headed north. La Salle later returned to France, where he was authorized to set up a colony at the mouth of the Mississippi. However, his party of 320 could not find the river and landed in Texas, where many of them died or deserted. In 1687, La Salle was murdered by his men while still seeking the Mississippi.

OPPOSITE TOP Sieur de La Salle enters a village of Cenis Indians in May 1686, while searching for the Mississippi delta. His lack of success led to his murder by his disgruntled followers.

NOT GOOD NEIGHBOURS

Cartier's difficulties encouraged French colonists to turn their attention south. In 1562, Jean Ribault set up a colony of French Huguenots at Port Royal, South Carolina. Two years later the settlers abandoned the site, but René de Laudonnière led a second expedition of Protestants fleeing religious persecution, and built Fort Caroline on St John's River. However, with the backing of the Spanish crown, Pedro Menéndez de Avilés, governor of Florida and founder of St Augustine – the oldest European settlement in North America – attacked Fort Caroline and massacred all the "heretical" settlers.

LEFT René de Laudonnière and Athore, a local chief, meet at one of several columns erected by Jean Ribault.

THE SPANISH IN LATIN AMERICA

After Columbus's successful voyages, other Spanish mariners also sailed west to make their fortunes. In 1499, Alonso de Ojeda and Juan de la Cosa – both of whom had accompanied Columbus – reached the coast of Guiana and followed it north to present-day Colombia. Seeing houses built on stilts in the water, Ojeda named the land Venezuela, meaning "Little Venice". In 1500, Vicente Pinzón, who had commanded *Niña* on Columbus's first voyage, struck the coast of Brazil near modern Recife and explored the mouth of the Amazon, which he mistakenly identified as the Ganges. During the next decade, Juan Díaz de Solis, who had accompanied Martin Pinzón in circumnavigating Cuba, claimed Uruguay for Spain and discovered the River Plate.

OPPOSITE TOP Pizarro seizes the unsuspecting Atahualpa in Cajamarca. The priest, Father Valverde, who had been ignored when telling Atahualpa to accept Christianity, holds his crucifix aloft triumphantly.

OPPOSITE BOTTOM above This type of house – a thatched roof and sides over frames of poles, all atop wooden piles – led Alonso de Ojeda to name the area "Venezuela" meaning "Little Venice".

Another explorer, who made unsubstantiated claims both of accompanying Columbus and leading his own expedition in 1497, was Italian-born Amerigo Vespucci. What is certain is that Vespucci accompanied Ojeda and de la Cosa across the Atlantic; he then went south while they investigated to the north. Vespucci supposedly made several more voyages, perhaps discovering the harbour of Rio de Janeiro, possibly reaching the River Plate, and certainly visiting the Gulf of Darien. Regardless of the truth of Vespucci's claims, in 1507 the German Martin Waldseemüller produced maps that suggested these coasts were part of a new continent, which he named "America" in honour of Vespucci.

In 1509, Ojeda attempted to establish a settlement at San Sebastian on the coast of Colombia. Although it failed, the survivors and some newcomers were led by Vasco Núñez de Balboa – a stowaway fleeing debts in Hispaniola – to a new site at Darien. From there, Balboa gradually extended his control, and in 1513 he forced his way across the Isthmus of Panama, becoming the first European to see the eastern side of the Pacific Ocean, which he named the Southern Ocean. Balboa had less success politically, and his conflicts ended with his public beheading in 1519 after being arrested by a ruthless, illiterate, battle-hardened conquistador who had previously served under him: Francisco Pizarro.

Several years later, Pizarro and his colleague Diego de Almagro, having heard rumours of cities of gold to the south, began a series of expeditions towards what was called Peru. In 1531, Pizarro, his three half-brothers and some 180 men established a base near the empire of the Incas. The next year, in Cajamarca, Pizarro seized Atahualpa, the Inca emperor, held him until an enormous ransom had been paid, and then executed him by garrotte. Joined by reinforcements under Almagro, Pizarro then marched on the Inca capital Cuzco, which they sacked, and then began consolidating their hold on Peru.

In the years that followed, Pizarro founded the city of Lima, which became capital of the Spanish territory, and divided power among his brothers. Diego de Almagro was given land to the south, but became dissatisfied when he found little gold after marching through the southern Andes and back via the Atacama Desert. Upon his return he seized Cuzco, which led to conflict with Hernando Pizarro, who captured and executed Almagro. The supporters of Almagro took their revenge by assassinating Francisco Pizarro in Lima in 1541.

The year before his death, Pizarro sent his brother Gonzalo to explore the interior of their domain. Crossing the Andes in 1541, the party reached a tributary of the Amazon eight months later. But they found scant food in the midst of swamp and forest, so Gonzalo sent a group of men under Francisco de Orellana downriver in a barge to find supplies. When Orellana failed to return, Gonzalo made a tortuous retreat to Quito. Orellana, meanwhile, could not travel upstream against the strong current, and so continued downriver into the Amazon, following it for approximately 4,830 kilometres (3,000 miles) before reaching the Atlantic in August 1542. Three years later, Orellana attempted to explore more of his mighty river, but the expedition ended in disaster, including his death.

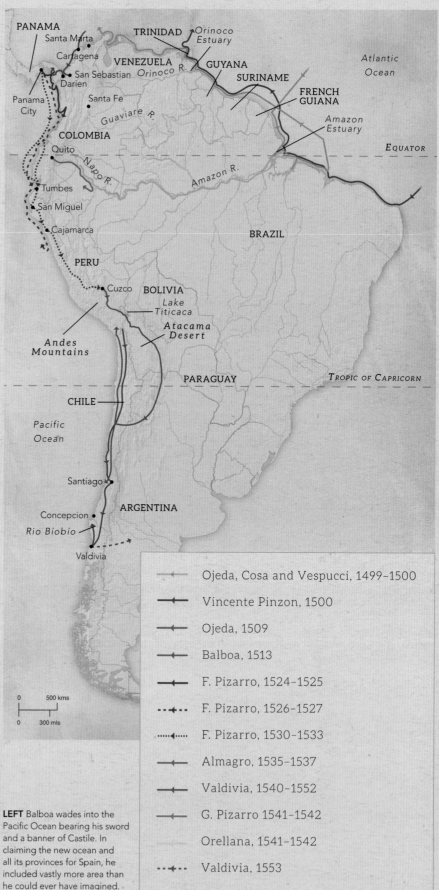

LEFT Balboa wades into the Pacific Ocean bearing his sword and a banner of Castile. In claiming the new ocean and all its provinces for Spain, he included vastly more area than he could ever have imagined.

→	Ojeda, Cosa and Vespucci, 1499–1500
→	Vincente Pinzon, 1500
→	Ojeda, 1509
→	Balboa, 1513
→	F. Pizarro, 1524–1525
--◄--	F. Pizarro, 1526–1527
····◄····	F. Pizarro, 1530–1533
→	Almagro, 1535–1537
→	Valdivia, 1540–1552
→	G. Pizarro 1541–1542
→	Orellana, 1541–1542
--◄--	Valdivia, 1553

THE TREATY OF TORDESILLAS

The voyages of Bartolomeu Dias and Columbus appeared to open a new area of competition between Spain and Portugal. In 1493, Pope Alexander VI issued a decree to forestall conflict between the two Catholic nations by dividing the unknown lands of the world between them. The next year, the Treaty of Tordesillas confirmed the arrangement, a line 370 leagues west of the Cape Verde Islands would separate Spanish territories to the west and Portuguese to the east; undiscovered Brazil lay in the Portuguese sector. The Treaty also antagonized the French, Dutch and English.

LEFT Part of the Treaty of Tordesillas, which in 1494 divided new discoveries throughout the world between Spain and Portugal, following Pope Alexander VI's previous decrees. The Treaty was even cited some 500 years later as validation for Argentina's claim to the Falkland Islands. (See Translations, page 204–205.)

THE CONQUEST OF CHILE

After Almagro's execution, his territories were given to Pedro de Valdivia, who had participated in the conquest of Venezuela before loyally serving Pizarro. Beginning in 1540, Valdivia made his way south through Chile, crossing mountains and deserts, drawing up plans for the city of Santiago in 1541, and founding Concepción in 1550. He then crossed the Rio Bíobío, but progress was halted in 1553 by the Araucanian Indians, who killed Valdivia and wiped out his army. The Araucanians maintained control over the area for nearly three centuries.

RIGHT Pedro de Valdivia experienced constant resistance in founding Santiago. Having established the settlement he proceeded to Valparaiso, but was forced to return to Santiago to quell a mutiny. While away, his new seaside garrison of Valparaiso was wiped out, and when Valdivia headed back to help, Santiago was attacked and destroyed.

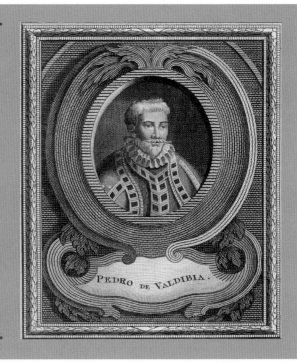

PEDRO DE VALDIBIA.

THE AMERICAS

ÐÑ̄ɪ ·1577·
Ǽŧaŧɪs sǽ·39·
Œ·ꜰ·

MARTÍN FROBISER Mɪlɪꜱ

·· ꝰᴇ ·

GᴠALTERɪ CHARLETON, M·D·

FROM HUDSON BAY TO BERING STRAIT

In the wake of Columbus's discoveries, other navigators soon began plying the western waters of the Atlantic. The first to approach the more northerly shores of the Americas was John Cabot, a Venetian who had settled in the English port of Bristol. In 1497, financed by a syndicate of merchants, Cabot sailed with a crew of just 18 on the tiny *Matthew*, making landfall in what was most likely Newfoundland. The next year he went west again, but he and four of his ships simply vanished. A decade later, Cabot's son Sebastian, who claimed to have accompanied his father's first expedition, attempted to find the Northwest Passage a sea route to Asia around the north of the North American continent but was stopped by ice on the Labrador coast.

Although the Portuguese brothers Gaspar and Miguel Corte-Réal independently reached the coast of Canada in the early sixteenth century – they were looking for a passage to Asia – it was the English for whom the search for the Northwest Passage became a national obsession. Throughout the sixteenth and early seventeenth centuries, numerous efforts were made to discover such a route, but all were unsuccessful because they were too far south. Martin Frobisher made three voyages (1576, 1577 and 1578): he discovered Frobisher Bay and encountered Inuit on the first; brought back hundreds of tons of what he thought was gold – but proved to be worthless ore – on the second; and discovered the entry to Hudson Strait on the third.

A decade after Frobisher, John Davis also made three voyages in search of the Northwest Passage. In 1585, he sailed up the coast of Greenland, then crossed Davis Strait to Baffin Island, where he discovered Cumberland Sound. The second voyage was similar to the first, but on the third, in 1587, he reached 72° 46' N before being stopped by ice.

The English search was resumed 20 years later by Henry Hudson, whose achievements helped lead to the formation of the Company of the Merchants Discoverers of the Northwest Passage, which subsequently financed many expeditions. The first, in 1612, under Sir Thomas Button, further explored Hudson Bay, and two others led by Robert Bylot (Hudson's mutinous mate) and piloted by William Baffin explored Baffin Bay and discovered its three major outlets, Smith, Jones and Lancaster sounds. Lancaster Sound was later recognized as the entrance to the Northwest Passage.

More than half a century later, the continuing search was formalized as part of a charter that King

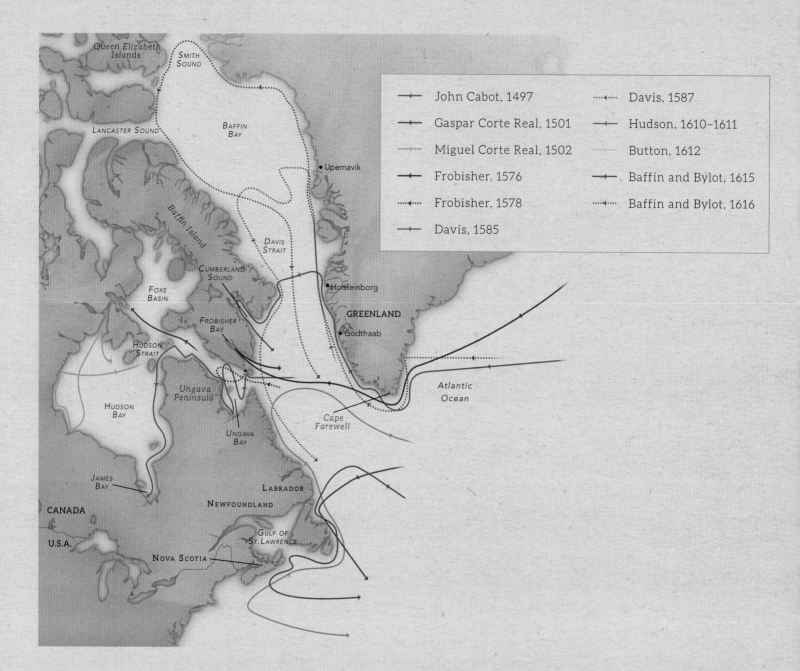

SMITH
SOUND

BAFFIN
BAY

LANCASTER SOUND

• Upernavik

Baffin Island

DAVIS
STRAIT

⟵ John Cabot, 1497	⟵ Davis, 1587
⟵ Gaspar Corte Real, 1501	⟵ Hudson, 1610–1611
⟵ Miguel Corte Real, 1502	⟵ Button, 1612
⟵ Frobisher, 1576	⟵ Baffin and Bylot, 1615
⟵ Frobisher, 1578	⟵ Baffin and Bylot, 1616
⟵ Davis, 1585	

CUMBERLAND
SOUND

FOXE
BASIN

• Holsteinborg

FROBISHER
BAY

HUDSON
STRAIT

GREENLAND

• Godthaab

Ungava
Peninsula

Atlantic
Ocean

HUDSON
BAY

Cape
Farewell

UNGAVA
BAY

JAMES
BAY

LABRADOR

NEWFOUNDLAND

CANADA

U.S.A.

GULF OF
ST. LAWRENCE

NOVA SCOTIA

Charles II granted to the "Governor and Company of Adventurers of England trading into Hudson's Bay", or the Hudson's Bay Company. This charter required ongoing exploration, while giving the Company exclusive economic control – including the lucrative fur trade, and mineral and fishing rights – over the entire drainage area of Hudson Bay, a vast territory known as "Rupert's Land". Despite later challenges from other mercantile competitors, including the North West Company, the Hudson's Bay Company only surrendered its rights to most of the territory in 1870, ceding them to the new Dominion of Canada.

Similar patterns of exploration and exploitation occurred in the north west of the continent, although at a much later date. Alaska was sighted in 1732 by the Russian Mikhail Gvozdev, but little importance

was attributed to the discovery. However, after Bering's Great Northern Expedition – a Russian effort to determine the geographical relationship between Asia and America (1733–43) – the exploration of the Aleutians and other coastal regions led to the growth of a lucrative fur trade. In 1799, the Russian government granted the Russian-American Company the exclusive use of all hunting grounds in Russian America, as well as the right to make discoveries and occupy newly discovered lands. The Russian hold on Alaska was strengthened by a series of Russian-American Company supply voyages, as well as the circumnavigation of the world under the command of Otto von Kotzebue (1815–18), which included extensive scientific investigation of Kamchatka, the Aleutians, Bering Strait and mainland Alaska.

OPPOSITE BOTTOM Sea ice such as this in the Victoria Strait in the Canadian Arctic is navigable, but it becomes less so as the ice floes become thicker and cover more of the surface, with fewer leads of water between them.

VANCOUVER'S SURVEY

In 1791, George Vancouver, a former midshipman under James Cook, was sent to explore a series of deep inlets in the north-west coast of North America. He was to discover if any were trans-continental passages through the Americas, as reported by Juan de Fuca in the 1590s. Over the next four years, Vancouver conducted a meticulous coastal survey from Washington to Alaska, including circumnavigating and charting the island that now bears his name. His survey finally laid to rest all hopes of a low-latitude Northwest Passage.

ABOVE George Vancouver, whose survey of the west coast of North America was one of the most thorough and detailed in the history of maritime exploration.

HENRY HUDSON

In 1607, Henry Hudson was commissioned by the Muscovy Company to search for a passage to the Orient via the North Pole. He sailed up the coast of Greenland and thence to Spitsbergen before being stopped by ice. The next year, Hudson made an attempt on the Northeast Passage, but was halted by ice near Novaya Zemlya. In 1609, after again being forced back near Novaya Zemlya, he sailed west to Nova Scotia, then followed the coast south, discovering the Hudson River, which he traced for 240 kilometres (150 miles) before realizing it had no northern outlet. In 1610, Hudson made his fourth voyage, discovering Hudson Bay. However, after spending a miserable winter near James Bay, some of the men mutinied, and Hudson, his son and seven others were set adrift in a boat, never to be seen again.zand massacred all the "heretical" settlers.

BELOW A romanticised image of Henry Hudson landing on Manhattan, where New York City is today located. In 1609 he discovered and explored the Hudson River.

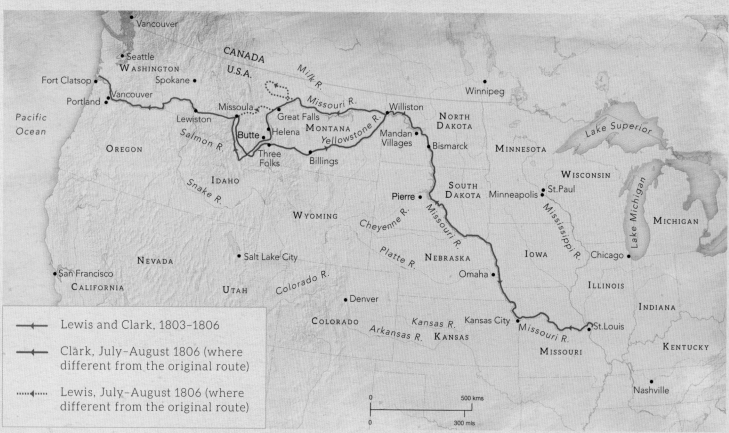

Vancouver

Seattle
WASHINGTON
CANADA
U.S.A.
Milk R.

Fort Clatsop
Vancouver
Spokane
Missoula
Great Falls
Missouri R.
Williston
Winnipeg

Portland
Lewiston
Helena
MONTANA
Yellowstone R.
Mandan Villages
NORTH DAKOTA
Lake Superior

Pacific Ocean
OREGON
Salmon R.
Butte
Three Folks
Billings
Bismarck
MINNESOTA

Snake R.
IDAHO
WISCONSIN
Minneapolis
St.Paul
Mississippi R.
MICHIGAN

WYOMING
Cheyenne R.
Pierre
SOUTH DAKOTA
Lake Michigan

NEVADA
Salt Lake City
Platte R.
Missouri R.
NEBRASKA
IOWA
Chicago
ILLINOIS

San Francisco
CALIFORNIA
UTAH
Colorado R.
Denver
Omaha

COLORADO
Kansas R.
Kansas City
Missouri R.
St.Louis
INDIANA

Arkansas R.
KANSAS
MISSOURI
KENTUCKY

Nashville

Lewis and Clark, 1803–1806

Clark, July–August 1806 (where different from the original route)

Lewis, July–August 1806 (where different from the original route)

0   500 kms

0   300 mls

LEWIS AND CLARK

One of Thomas Jefferson's goals when he became the third president of the United States in 1801 was to establish the young nation's economic independence, in part by trading throughout and beyond the huge area of the North American west. Before the Louisiana Purchase was even finalized, Jefferson persuaded Congress to grant money for an expedition that would determine whether communication and trade across the continent were possible via the Missouri river system and its possible links to already discovered rivers that ran into the Pacific Ocean.

Placed in charge of the expedition was Jefferson's 29-year-old personal secretary, Meriwether Lewis, who then chose a former fellow army officer, William Clark, aged 33, as his co-commander. In addition to geographical information, the two were also asked to compile detailed scientific observations of the rivers, terrain, natural resources, weather and native peoples of the lands traversed.

In late 1803, Lewis and Clark assembled a party of about 45 men, and wintered across the Mississippi River from St Louis, waiting for the spring – and for the local French authorities to learn of the Louisiana Purchase – before beginning their ascent of the Missouri River. In May 1804, the Corps of Discovery, as the expedition was known, set off in a 17-metre (55-foot) keelboat and two small sailing boats called pirogues, accompanied by horsemen on the banks. With strict instructions about maintaining friendly relations with the Indians, the party avoided any conflicts as it moved slowly northwest through prairies dominated by vast herds of buffalo. They wintered at a Mandan village in what is now North Dakota, where they were joined by the French-Canadian fur trader Toussaint Charbonneau and his Shoshoni wife Sacagawea.

In April 1805, Lewis and Clark moved ahead in six canoes and the two pirogues, making good time past the confluence of the Missouri and Yellowstone rivers. In June, they reached the Great Falls of the Missouri (in what became Montana), where they spent three weeks portaging to where they could take to the river again. Then they ascended the foothills and trekked into the Rocky Mountains, suffering from hauling their equipment in the burning heat of the day while spending nights in bitter cold. Having passed through the heart of the Rockies, they were faced not with a broad plain leading towards the sea, but instead with the peaks of the Bitterroot Mountains. These, too, they struggled through, abandoning their canoes on the way, and on the far side meeting a band of Shoshoni led, remarkably, by Sacagawea's brother.

After spending several weeks with the Shoshoni – who provided information about the way west and sold them horses – the expedition moved on at the end of August. They made friends with a band of Nez Percé Indians and, building more canoes, descended the Clearwater River to the Snake and then the Columbia river. They reached the Pacific Ocean in November, having travelled, according to Clark, 6,791 kilometres (4,192 miles).

OPPOSITE TOP LEFT & RIGHT
The leaders of the Corps of Discovery were William Clark (left) and Meriwether Lewis (right). They were both Virginians from good families who had first met in 1795 while serving in the army.

The Corps of Discovery spent a wet, miserable winter near the coast before, in March 1806, heading home. In July, Lewis and Clark divided the party, with Lewis making a diversion north of the original route and Clark following the Yellowstone River south of it. When one of the party, John Colter, resigned from the expedition on the return journey to go and trap beaver, he became the first white man to see the geysers of what became Yellowstone National Park, although, when he told his tale, no one believed him. The two parties met up again in mid-August where the Yellowstone met the Missouri and raced downriver on a fast current, reaching St Louis in September.

Lewis and Clark had not found a waterway across North America, but they had explored and mapped huge swathes of unknown territory, documenting new lands, new specimens of flora and fauna and the languages and customs of numerous Indian tribes. They had begun the process of opening up the west to future settlers and to its eventual incorporation into the United States.

ABOVE Lewis and Clark came into contact with many different Native American tribes. This drawing first appeared in the journal of Peter Gass, an expedition member.

LEFT A canteen from the time of the Lewis and Clark expedition. This would have been vital for those attempting to cross the great western deserts.

OPPOSITE TOP RIGHT Travelling by water along the Missouri River was difficult and dangerous, and required frequent trips to shore for transporting food, arms, powder, scientific instruments, goods to be bartered and the boats themselves.

OPPOSITE BOTTOM A map produced following the journey of Meriwether Lewis and William Clark across the territories of the Louisiana Purchase and other far-distant parts of North America. Never before had the public known the geography and the natural marvels of these regions.

SACAGAWEA

Born in 1786 or 1788, Sacagawea was a Lemhi Shoshoni. She was captured by the Hidatsas when she was about 13 and later sold to the French-Canadian fur trader, Toussaint Charbonneau. At the Mandan village, Charbonneau joined Lewis and Clark as an interpreter and took along Sacagawea and their two-month-old baby. Although legend suggests that Sacagawea was the expedition's guide, this overstates her role. In reality, she served as another interpreter and her presence with a baby reassured the native peoples that the expedition had friendly intentions. She died around 1812 at Fort Manuel in South Dakota.

LEFT Some legends indicate that Sacagawea was the key guide for the Corps of Discovery. Although these probably give her role too great an importance, her place in history is secure.

THE LOUISIANA PURCHASE

The political process that culminated in the Louisiana Purchase had actually begun 40 years previously, when, at the close of the Seven Years' War, Spain took control of the French lands west of the Mississippi River. However, by the Treaty of San Ildefonso in 1800, the Spanish returned Louisiana to France. Napoleon, needing capital for his European wars, thereupon sold the vast territory to the United States in 1803. For a payment of $27,267,622, Thomas Jefferson's government obtained around 2,144,250 km2 (828,000 square miles), thereby doubling the size of the US.

RIGHT The Louisiana Purchase, which was bound in black leather to protect it. The signing of the Purchase in 1803 effectively doubled the size of the US.

MANIFEST DESTINY

Thomas Jefferson, like many early Americans, believed his new nation was destined to explore, settle and control the territories extending to the Pacific Ocean. The Louisiana Purchase and Jefferson's backing of Lewis and Clark were key factors in maintaining western expansion as a central vision of the American republic. In 1845, John O'Sullivan, editor of The United States Magazine and Democratic Review, labelled this belief in a national mission "Manifest Destiny". This innate conviction lasted throughout the nineteenth century, with fur traders, military men and professional explorers helping to open up vast new areas.

OPPOSITE TOP The Gateway to the Garden of the Gods in Colorado, leading to snow-capped Pikes Peak. Zebulon Pike wrote that the "Grand Peak", which he estimated to be 18,000 feet, would likely never be climbed. However, in 1820 Edwin James reached the summit, which is actually 14,110 feet. In 1893, Katherine Lee Bates was so inspired by the view from the top of Pikes Peak that she composed the lyrics to "America the Beautiful".

OPPOSITE BOTTOM LEFT Jedediah Smith, pictured here fighting a bear, had more than his share of death-defying adventures. In 1828 Smith and two companions returned from reconnoitring ahead to find Indians had massacred everybody who had been left in camp.

OPPOSITE BOTTOM RIGHT Only in his twenties when his expeditions made him a national figure, Zebulon Pike returned to military duties and in 1812 was promoted to Deputy Quartermaster General. During the War of 1812 he was killed in an assault on the Canadian town of York (now Toronto).

In 1805, immediately after Lewis and Clark's expedition, the governor of the Louisiana Territory sent an army officer, Zebulon Pike, to reconnoitre the headwaters of the Mississippi River and report on the British fur trade there. The next year, Pike received secret instructions to spy on the Spanish in New Mexico. He headed west to the Rocky Mountains, where he sighted the 4,300-metre (14,110-foot) Pikes Peak in Colorado. His party then moved south to the Rio Grande where, in 1807, Pike was imprisoned for three months by the Spanish, during which time he made extensive studies of their frontier defences. His reports and published account contained detailed information on both the Spanish territories and the Central Plains.

The year after Pike returned, John Jacob Astor, a former emigrant from Germany, founded the American Fur Company, which would eventually break the British-Canadian dominance in the trade. Astor sponsored numerous expeditions to open the northwest, and his company's trappers made many discoveries. Perhaps the most important expedition was by Robert Stuart, who in 1812 made the first documented crossing of the Rocky Mountains at South Pass, establishing what would become the Oregon Trail.

Other fur traders followed, many in the employ of William Henry Ashley. Ashley himself led expeditions to Wyoming's Green River and later along the Platte River, through the Rocky Mountains, to the Great Salt Lake. In 1826, he sold his company to three associates, including the greatest of exploring trappers, Jedediah Smith. Smith made three major journeys. The first explored the Rocky Mountains north into Canada. Then, in 1826–27, Smith left the Great Salt Lake, crossed the barrens of Utah and northern Arizona, and blazed a trail through the Mojave Desert and the San Bernardino Mountains, eventually reaching Los Angeles. After suffering house arrest by the Spanish, he travelled north through the San Joaquin Valley, then made the first west–east crossing of the Sierra Nevadas. Smith's third major expedition, in 1827–28, once again crossed the Black Mountains and the Mojave. This time, upon reaching the coast he headed north, passed San Francisco, and continued to the Hudson's Bay Company post on the Columbia River before making his way southwest through the Rockies. There were

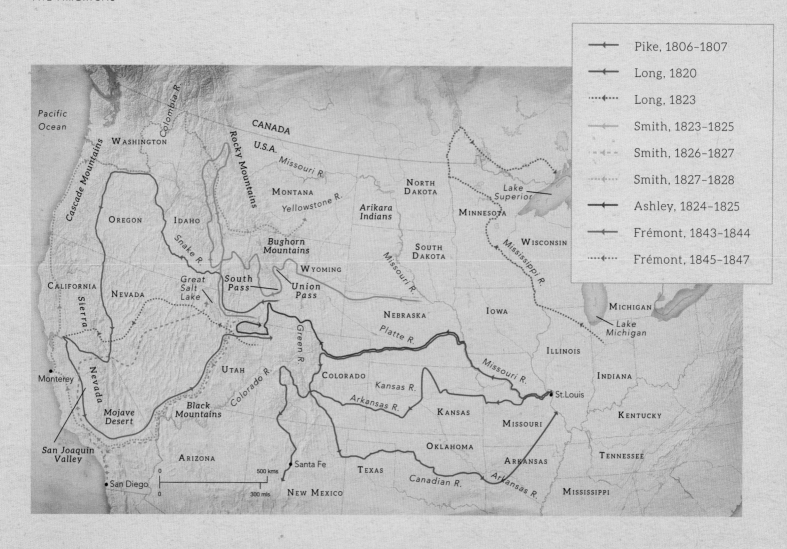

Pike, 1806–1807
Long, 1820
Long, 1823
Smith, 1823–1825
Smith, 1826–1827
Smith, 1827–1828
Ashley, 1824–1825
Frémont, 1843–1844
Frémont, 1845–1847

JOHN WESLEY POWELL

Despite the loss of his right arm at the battle of Shiloh, John Wesley Powell became a great explorer. After preliminary investigations with his wife, in 1869 Powell launched a nine-man, 2,400-kilometre (1,500-mile) trip down the Green and Colorado rivers. For more than three months they pitched over falls, struck massive boulders and shot rapids, making the first expedition through the Grand Canyon and exploring the last major unknown region of the 48 contiguous states of the United States. In 1872, Powell returned to the American southwest, conducting a topographical survey of a huge area of previously uncharted territory.

LEFT John Wesley Powell (far left) with three other members of his remarkable first expedition.

numerous other legendary "Mountain Men" besides Smith, including Jim Bridger, Joseph Walker, Kit Carson and Jesse Chisholm.

Like fur traders, members of the US Army Corps of Topographical Engineers also helped open the way west. In a series of river-oriented expeditions, Stephen Long followed sections of the Mississippi, the Arkansas, the Missouri (becoming one of the first men to travel on it in steamboats) and the Platte. He was also in charge of the first ascent of Pike's Peak and produced maps that called the plains the "Great American Desert", thereby helping discourage settlement of that region for several decades.

Another officer in the Corps of Topographical Engineers was John C. Frémont. In the early 1840s, he commanded two official expeditions into the Oregon Territory – an area also claimed by Britain – continuing on the second into Spanish California. Frémont's two published expedition accounts, written with his wife, made him and his guide, Kit Carson, national legends. But indulging in politics while he was supposed to be concentrating on a third expedition led to his dismissal from the army. His reputation remained so strong that in 1856 he became the presidential candidate of the new Republican Party, although he lost the election to James Buchanan.

BELOW In 1843 the first large-scale migration of settlers along the Oregon Trail reached the Willamette Valley. Within 15 years, more than a quarter of a million people had made the six-month trek from near Independence, Missouri along the Platte River to Fort Laramie, through the Rocky Mountains, and finally through the Blue Mountains to the Columbia River.

THE FIRST PROFESSIONAL EXPLORER

Frederick Schwatka held both medical and legal qualifications, but instead became an explorer. He first earned a reputation on the American Geographical Society's Franklin search (1878–80), in which he made a record sledge journey of 5,232 kilometres (3,251 miles). In 1883, he made a reconnaissance of Alaska, including a remarkable combination overland and river journey. Schwatka then "turned pro", willing to travel anywhere for sufficient payment. In the following decade, he unsuccessfully attempted to climb Mount St Elias and make the first winter crossing of Yellowstone, but did explore the Sierra Madre Mountains, and follow the Copper River.

LEFT No organizations were more involved in funding Frederick Schwatka than the New York newspapers. His first expedition was backed by *The Herald*, and later ones by *The Times*, *The World* and *The Ledger*.

THE SCIENTIFIC EXPLORATION OF SOUTH AMERICA

By the late seventeenth century, much of South America had been penetrated by conquistadors, missionaries, or slave-traders. But the first comprehensive explorations of the hinterlands, native peoples, wildlife and flora did not begin until the mid-eighteenth century, when numerous scientific expeditions were launched.

OPPOSITE An illustration from one of von Humboldt's books, showing him, Aimé Bonpland, and their guide, with the background dominated by Mount Chimborazo which, at the time, was thought to be the world's highest mountain. Their ascent to more than 19,000 ft (5,600 m) was the highest that had been accomplished to that time.

ABOVE A bust of Charles-Marie de la Condamine. His Amazon journey took longer than expected because he spent much time taking scientific measurements and observations. His map of the Amazon remained relatively unimproved upon until the twentieth century.

One of these was designed to determine the precise shape of the Earth, and whether it flattened near the Poles, as Isaac Newton predicted, or towards the Equator. In 1735, the Frenchman Charles-Marie de La Condamine was sent to Quito to measure the exact length of a degree of latitude at the Equator, while similar measurements were made in Lapland. For seven years, La Condamine continued his investigations, until hearing that the party in Lapland had shown Newton to be correct. Instead of returning directly to France, La Condamine decided to trace the Amazon to the Atlantic, as Orellana had done two centuries before. His journey, in 1743, was the first such expedition led by trained scientists.

Meanwhile, one of La Condamine's scientists, Jean Godin des Odonais, remained in Quito, having married a Peruvian woman. In 1749, called home because of his father's death, Godin followed La Condamine's route, leaving his pregnant wife Isabel behind because the journey was too difficult. It proved impossible to return to collect her, and it was only 20 years later that Isabel was told that a ship was waiting for her on the upper Amazon. She then followed her husband's route, accompanied by three kinsmen, three Frenchmen and 35 servants and porters. High on the Amazon,

however, the porters deserted and, after one of the Frenchmen drowned, the others went for help – never to be seen again. Isabel waited with her family for 25 days before setting off on foot. They became lost in the forest, and everyone died except Isabel, who lived off roots and insects before being rescued by locals. She eventually made her way downriver and joined her husband in Guiana.

The turn of the nineteenth century saw the arrival of one of the world's most prodigious scholars, Baron Alexander von Humboldt. For five years, he and French botanist Aimé Bonpland explored vast regions and made astonishing contributions to contemporary science, including collecting more than 6,000 previously unknown plant species; charting and making hydrographic investigations of numerous rivers; conducting studies of electricity and establishing the existence of the current that flows north along the South American coast and that bears Humboldt's name today. Their ascent of Mount Chimborazo set an altitude record that was not surpassed for many years.

In the following decades, other scientists uncovered yet more of the interior. The Brazilian rainforest attracted the English collector William Burchell,

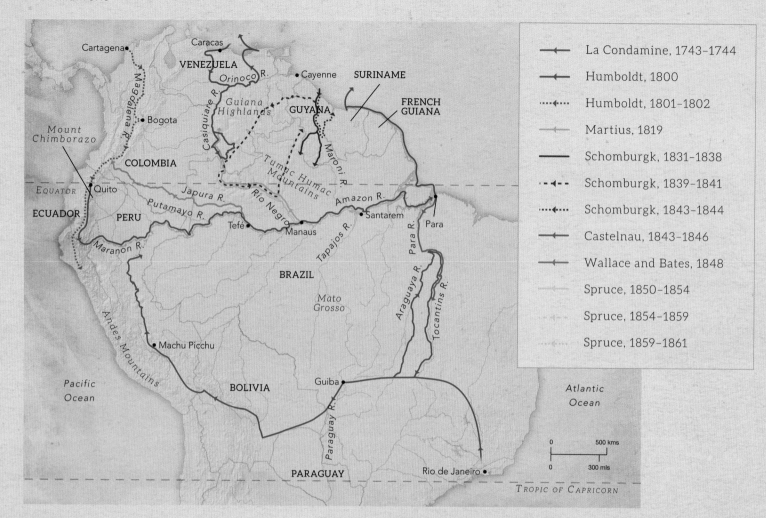

Legend:
- La Condamine, 1743–1744
- Humboldt, 1800
- Humboldt, 1801–1802
- Martius, 1819
- Schomburgk, 1831–1838
- Schomburgk, 1839–1841
- Schomburgk, 1843–1844
- Castelnau, 1843–1846
- Wallace and Bates, 1848
- Spruce, 1850–1854
- Spruce, 1854–1859
- Spruce, 1859–1861

while the German naturalist Georg von Langsdorff surveyed the Mato Grosso region. The Frenchman Auguste de Saint-Hilaire wandered through Brazil for five years, covering some 11,300 kilometres (7,000 miles), and Philipp von Martius and Johann von Spix returned to Bavaria in 1820 with 6,500 plant and 3,300 animal specimens. For more than a decade, Robert Schomburgk undertook a wide range of investigations in British Guiana for the Royal Geographical Society. Meanwhile, a Swiss naturalist, Johann von Tschudi, investigated central Peru, and the French Count of Castelnau led an expedition from Rio de Janeiro that covered huge swathes of hitherto unexplored Brazil and Paraguay.

None were more successful, however, than a pair of Britons who set off in 1848. Henry Walter Bates and Alfred Russel Wallace journeyed thousands of miles together and separately before, in 1852, Wallace departed for England. Bates continued collecting until 1859, when he returned to England with more than 14,000 species of insects, including some 8,000 new to science. Wallace was less fortunate; while on

THE LOST INCA CITIES

After experiencing Francisco Pizarro's brutal treatment, the last Inca rulers withdrew to refuges in the Andes, where they, too, were eventually exterminated. Nearly four centuries later, in 1911 a Yale professor named Hiram Bingham launched an expedition to find the last Inca capital, Vilcabamba. High in the Andes, Bingham was shown what he thought was Vilcabamba, which he named Machu Picchu. He later found another magnificent site he called Espíritu Pampa. Half a century later, Gene Savoy demonstrated that it was not the spectacularly located Machu Picchu but Espíritu Pampa that was Vilcabamba.

ABOVE Machu Picchu, with its terraces, remains an incredible setting and seemed to leap from the mists of time when discovered by Hiram Bingham.

the way home, his ship caught fire. He and the other passengers were saved, but his entire collection was lost. Undaunted, he began an eight-year journey through Malaysia and Indonesia, during which he collected more than 127,000 specimens. In the course of this work, Wallace independently, and concurrently, reached similar conclusions to Charles Darwin on the process of evolution.

RIGHT Alexander von Humboldt, pictured in his library in 1856. One of the true giants in the history of science, he spent much of the 30 years following his return from South America producing some 35 volumes covering a wide range of scientific subjects based on his collections and observations.

His later years were devoted to Kosmos, his multi-volume treatise in which he attempted to provide a comprehensive picture of the universe and mankind's place in it.

BELOW A sketch map drawn by Alexander von Humboldt of part of the Orinoco River.

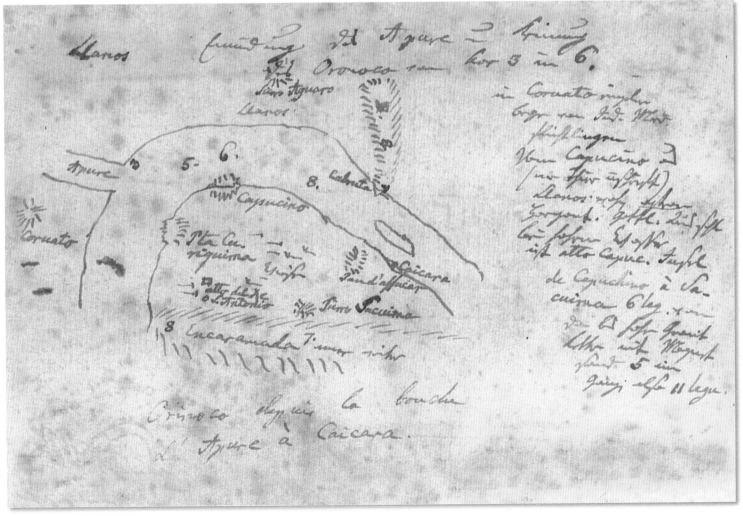

THE HUNT FOR THE CHINCHONA TREE

In the mid-nineteenth century, it became accepted that quinine, derived from the bark of the chinchona tree, protected against malaria. But chinchona were only found in the mountains of Ecuador, Bolivia and Peru, which maintained monopolies on its sale. In 1859, the India Company secretly assigned a clerk, Clements Markham, and the explorer and botanist Richard Spruce to gather seeds and saplings of chinchona and ship them to India for cultivation. Although Markham's samples were not successful, Spruce shipped thousands of seedlings, which eventually served as the basis for the chinchona plantations in Madras.

RIGHT Clements Markham first earned renown in the hunt for chinchona trees. He later became the president of the Royal Geographical Society and was a key driving force in kick-starting British Antarctic exploration, including selecting Robert Falcon Scott to lead the Discovery Expedition.

BELOW AND OPPOSITE A series of extracts from one of two of Clement Markham's notebooks compiled while searching in Peru for seeds and saplings of the chinchona tree.

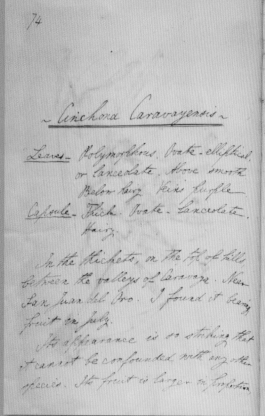

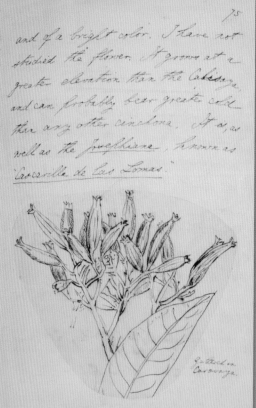

72

in April and May. Seeds ripen in
August and September.

Varieties   α Pelletieriana –
            β Purpurea –

α – Leaves on both sides green.
    Is called from Pelletier having
    discovered the alkaloid called
    aricine in its bark. It is
    known as Carua Carua or Cargua Cargua.
    Which means Llama, but figuratively
    "very bad." In Caravaya it is
    also called Cascarilla Amarilla.

β – Adult leaves purple beneath.
    Found near Huanuco.

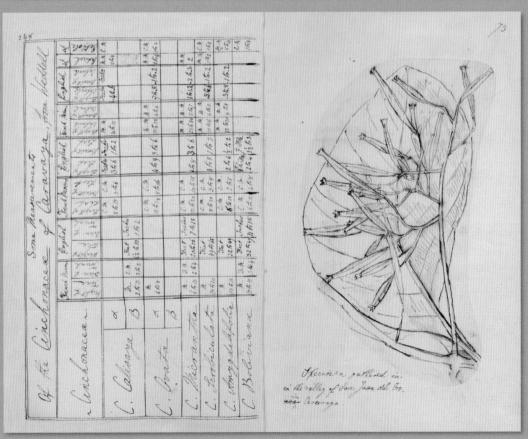

Specimen gathered in
in the valley of San Juan del Oro,
Caravaya.

ASIA

NIÂ

Abij Scythæ

Nubiæ lacus

Dtuidna fl

Colmogora
regio

Alani Scythæ

Auzacitis
regio

Aruibæ montes

Tapuri

Massagetæ

Oechardes

Rha fl.

Aspa fit montes

Iaxartes

Saccæ

Astoræ

Casita montes

Auzacia

Sogdiana

Bilthe

Cassia

Isfedon

Mare Caspiū

Oxus fl.

Emodij montes

MEDIA

Hyrcania

Bactri
ana

Rymus mons

INDIA

PERSIS

Itra Gãgem

PARTHIA

Aria

Arachosta

INDIA

Dragiana

citra Gãgem

Ganges

Sinus Persicus

Indus fl.

Carmania

Cambaia

Narsingæ
regnum

Regio Ar
gentea

Ormus

Sinus Guzerat

Goa

Pego

BIA

Canonor

Sinus Gangeticus

Regio aurea

Aden

Calicut

brum

Zaylon

Malaqua

Christiana

TAPRO
BANA

Mare
rassodum

Sumatra

Madagasiar

Zanzibar

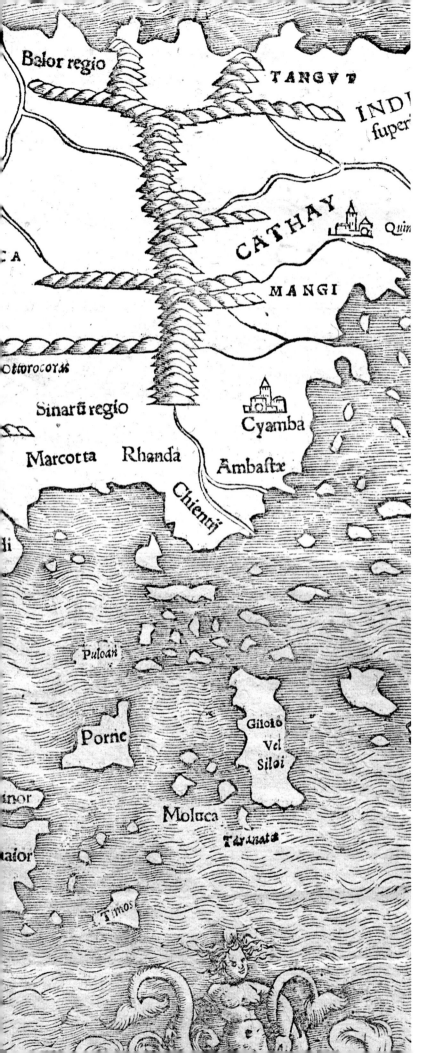

LEFT This is the earliest printed map of the Asian continent, largely based on contemporary geographical discoveries by Portuguese navigators. It was drawn around 1540 by Sebastian Münster, and appeared in his famous book *Cosmographia*, Germany's first description of the world. "India Extrema", as the map is known, shows Asia from the Caspian Sea and Persian Gulf to the Pacific. The outline of the mainland is reasonably well-established, with Java Minor and Major, Porne (Borneo), Moluca and several other islands named. The map does not include Japan and Korea.

EUROPEAN EXPLORATION OF CHINA AND SOUTHERN ASIA

Europeans frequently entered the New World, sub-Saharan Africa and the Arctic with beliefs in both their superiority to native peoples and the need to explore such "benighted" areas to expose them to European civilization. However, the cultures of China, India and other Asian lands were ancient, magnificent and clearly at least the equal of those Western countries, raising the question of whether such highly civilized areas could be "explored" at all. In fact, the Europeans who first visited those areas were not explorers in the regular sense, but rather merchants or missionaries.

Although the Italian friar Giovanni del Pian da Carpini and his French counterpart William of Rubruck had journeyed far across Asia into the domains of the Mongol Empire in the mid thirteenth century, it was Marco Polo's account entitled *Description of the World* that opened up to Europe the many wonders of what he called Cathay.

A century and a half later another Venetian, Nicolò de' Conti, spent 25 years travelling through Mesopotamia, Persia, India and Southeast Asia. Forced upon pain of death to convert to Islam, he later confessed to Pope Eugenius IV that he had temporarily abandoned Christianity. The Pope's forgiveness required only that Conti record an account of his wonderful journeys, which became another travel classic.

Two of the most circuitous penetrations into Asia came under the leadership of the Englishman Anthony Jenkinson. An envoy of the Muscovy Company, in May 1557 Jenkinson departed from Gravesend to follow the dangerous sea route pioneered by Richard Chancellor around Scandinavia, through Archangel'sk and south to Moscow. Wishing to open trade with Cathay, he then pushed south along the Volga River, across the Caspian Sea and east to Bukhara, where he arrived in December

ABOVE The Central Asian city of Bukhara was once the centre of a powerful khanate and still boasts magnificent architecture. It was visited by Ibn Battuta and Maffeo and Nicolò Polo before being subdued by Russia in the 1860s.

LEFT Anthony Jenkinson visited magnificent rulers in Aleppo, Moscow, Bukhara and Persia, in the hope of winning trading privileges for the Muscovy Company.

RIGHT A Persian dagger that belonged to James Silk Buckingham. After adventures including being press-ganged, stricken by plague, and robbed by bookkeepers in London and pirates aboard ship, Buckingham travelled to Oman, Egypt, Syria, Mesopotamia and Persia before becoming a newspaper editor, Member of Parliament, and social reformer.

UNVEILING THE MEKONG

In 1863, the French government declared Cambodia to be a protectorate. Three years later, an expedition was launched to determine if the Mekong River could be navigated, thereby creating a trade route to China. Ernest-Marc-Louis Doudart de Lagrée, the French representative to the Cambodian king, was placed in charge. After making the first comprehensive survey of the site of ancient Angkor, the expedition ascended the Mekong, eventually discovering the tumultuous Khong Falls, which eliminated any possibility of navigation. Doudart de Lagrée died in 1868, but the expedition continued into the Chinese province of Yunnan, the final mapping covering some 6,500 kilometres (4,000 miles).

RIGHT Along the Mekong River Doudart de Lagrée witnessed many local customs and competitions, such as these pirogue races that were illustrated in his posthumously published book.

FAR LEFT The Italian Jesuit Matteo Ricci went to China in 1582. He moved north, establishing missions in Nanchang and Nanking, before being allowed the rare honour of settling in Beijing in around 1600.

LEFT The rugged Pamirs, "The Roof of the World", separate Afghanistan and Pakistan from Tibet and western China.

BELOW A seventeenth-century engraving of the Potala in Lhasa, as it might have been seen by fathers Johann Grueber and Albert d'Orville when they became the first Europeans to reach the Tibetan capital.

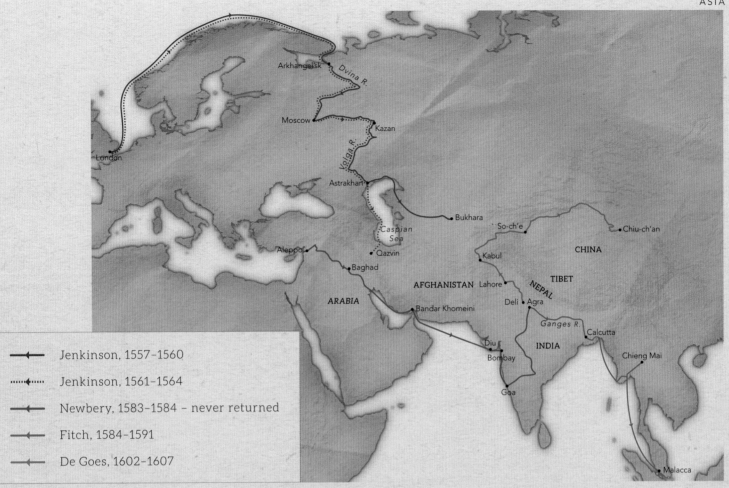

Jenkinson, 1557–1560

Jenkinson, 1561–1564

Newbery, 1583–1584 – never returned

Fitch, 1584–1591

De Goes, 1602–1607

1558. But trade with China via Bukhara had ceased, so Jenkinson retraced his course to England. In 1561, he began again, following the same route to the Caspian Sea but sailing to its southern end, from where he travelled into Persia. But the Persian shah did not want Christian traders in his domain, and Jenkinson was again forced to return by the same long, slow route.

The English Levant Company hoped to succeed where the Muscovy Company had not, and in 1583 sent merchants John Newbery and Ralph Fitch to open an overland route to India and beyond. Seen as a threat to Portuguese control, they were arrested in the Portuguese trading post of Hormuz and sent to Goa for trial. However, they escaped and made their way to the court of Akbar, the Mughal emperor, in Agra. There they split up; Newbery hoped to return to London overland, but he vanished on the way, never to be seen again. Fitch, meanwhile, sailed down the Ganges, visiting the Himalayas, Bengal, Burma, Siam and Malacca before reaching England in 1591, having been away eight years and long since given up as dead.

In the seventeenth century, Portuguese, Dutch and English commercial ventures in Asia concentrated on India and Indonesia, but inland Asia became the focus of a series of Jesuit priests, among the first being Matteo Ricci, who established missions between Canton and Beijing. One of Ricci's contemporaries was Benedict de Goes, originally from the Azores. In 1602–07, de Goes made a majestic sweep through central Asia, departing from Agra, visiting Lahore and Kabul, crossing the Pamirs and following the Silk Road through the Taklamakan and Gobi deserts, becoming the first European to travel overland into China since the fourteenth century. He died in Suchow in the spring of 1607.

The Jesuits next took aim at the mysterious land of Tibet, and in the 1620s Antonio Andrade left Agra and crossed the Himalayas to Tsaparang in western Tibet. The same decade, two more Jesuits, Estevão Cacella and João Cabral, reached the holy Tibetan city of Shigatse from Bhutan. Missions were set up in both settlements, although neither lasted long. The next great Jesuit journey came in 1661–62, when fathers Johann Grueber and Albert d'Orville left Beijing, crossed China, entered Tibet and became the first Europeans ever to reach the capital of Lhasa.

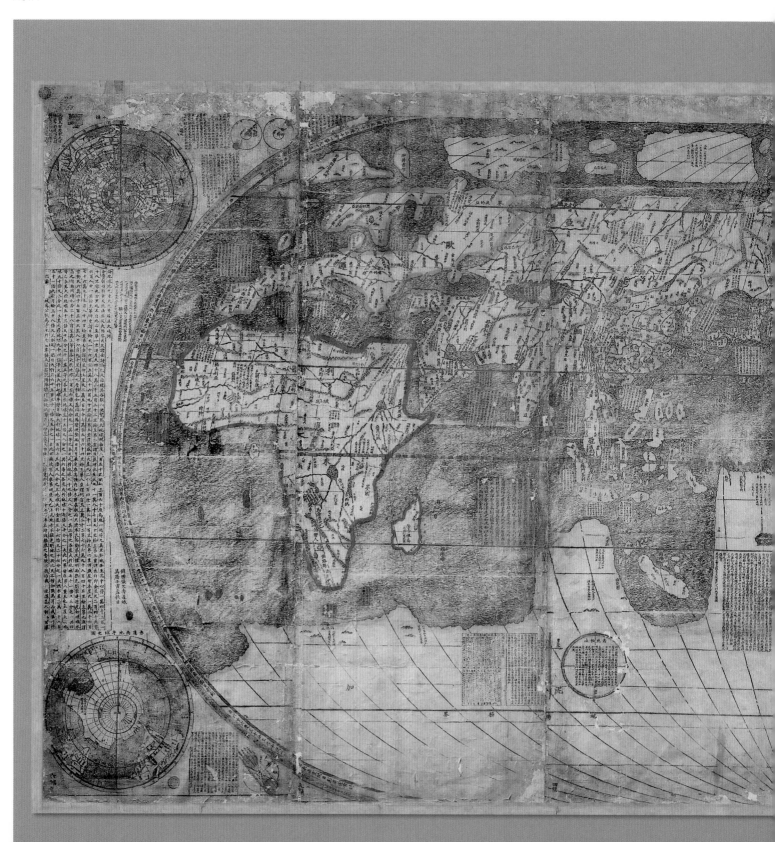

ABOVE In 1602, this Jesuit map of the world was printed in China. Note that, like a number of European maps (such as the famous one produced by Abraham Ortelius in 1570), this map shows a large Antarctic continent, which was then unknown but theorized to be there.

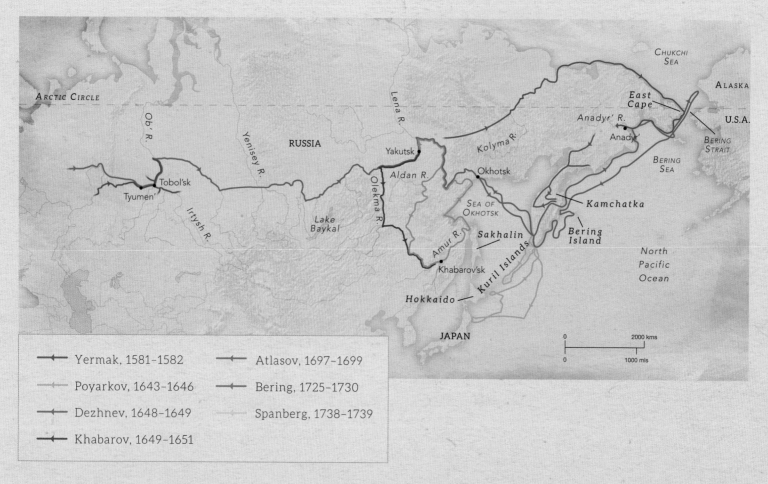

ARCTIC CIRCLE

CHUKCHI
SEA

ALASKA

RUSSIA

East
Cape

U.S.A.

Ob' R.

Yenisey R.

Lena R.

Kolyma R.

Anadyr' R.

Anadyr'

BERING
STRAIT

Tobol'sk

Yakutsk

Aldan R.

Okhotsk

BERING
SEA

Tyumen'

Irtysh R.

Lake
Baykal

Olekma R.

SEA OF
OKHOTSK

Kamchatka

Bering
Island

North
Pacific
Ocean

Amur R.

Sakhalin

Khabarov'sk

Kuril Islands

Hokkaido

JAPAN

0   2000 kms

0   1000 mis

Yermak, 1581–1582	Atlasov, 1697–1699
Poyarkov, 1643–1646	Bering, 1725–1730
Dezhnev, 1648–1649	Spanberg, 1738–1739
Khabarov, 1649–1651	

SIBERIA

Uniquely in the history of geographical discovery, Siberia was explored not from its coastlines inland, but from the interior to the periphery. This was because, whereas most of the coasts are ice-bound much of the year, the frozen Siberian plains and forests are dissected by navigable rivers that allowed explorers or conquerors to advance across the region. This happened despite the worst climate in the Northern Hemisphere and its seemingly endless size: 14 million km2 (5.3 million square miles) – roughly one and a half times the United States.

The penetration of Siberia began in 1581, after the powerful Stroganov family, which controlled a vast commercial empire west of the Urals, received rights from Ivan IV to lands east of those mountains. They initiated a campaign under the Cossack Yermak Timofeyevich against the Siberian Tatar khanate ruled by Kuchum. Enjoying a significant advantage in firearms, Yermak quickly won a series of battles and took Kuchum's capital near present-day Tobol'sk at the confluence of the Irtysh and Ob' rivers. Following this, Kuchum continued a nomadic, guerrilla-type resistance for more than 15 years. However, although Yermak was killed in 1585, systematic Russian expansion into Siberia had begun.

In rapid succession the Russians followed the system of rivers east and north, establishing military and trading outposts, including Tobol'sk, Tomsk on the Tom' and Turukhansk on the mighty Yenisey. In the 1620s, they pushed to the vast Lena River, where Petr Beketov founded Yakutsk in 1632. This served

as a launching point for future expeditions, and in 1643–46 Vasiliy Poyarkov went south to the Amur River, which, in the violent and murderous style of the conquistadors, he followed to the Sea of Okhotsk. Within several years, Yerofey Khabarov made two more expeditions to the Amur, demonstrating an equally bloodthirsty temperament, which led to skirmishes with the Chinese, who already claimed the area.

Farther north, Ivan Rebrov reached the Indigirka River in 1638, followed in 1643 by Mikhail Stadukhin's discovery of the easternmost of the great Siberian rivers, the Kolyma. In 1648–49, Semen Dezhnev set out from the Kolyma and sailed with seven ships along the Arctic coast. All were wrecked or disappeared, but the final one first passed through the Bering Strait, thereby discovering the easternmost extremity of Asia. However, the importance of this was not recognized, and by the time Vitus Bering repeated the act 80 years later, Dezhnev had been forgotten.

OPPOSITE This painting, "The conquest of Siberia by Yermak" by V. I. Surikov, was not produced until 1895, but it vividly captures a battle between Yermak's Cossacks and the Siberian Tatars more than 300 years before on the Irtysh River.

LEFT This famous depiction shows the death of Vitus Bering in December 1741 on Bering Island. Georg Steller effectively took command thereafter and in August 1742 led the 13-day voyage in the new ship back to Petropavlovsk.

BELOW LEFT A report written by Vitus Bering in 1740. Despite Bering having set out from St Petersburg in 1733, his ships, *Sv Petr* and *Sv Pavel*, were not finally launched until 1740. (See Translations, page 205.)

BELOW RIGHT The frozen coastline of Siberia extends for thousands of miles along the many seas to its north. The seaway to the north was once known as the Northeast Passage, but today is called the Northern Sea Route.

In 1725, Tsar Peter the Great sent Bering, a Dane in the Russian navy, to determine whether Asia was linked to North America. This was no simple task, as Bering's party had to cross Siberia, build a ship to take them across the Sea of Okhotsk to Kamchatka – which had been ruthlessly conquered by Vladimir Atlasov in the 1690s – cross that peninsula, build two more ships, and then sail north between the continents. All of this Bering achieved in five years, but when he returned, others were sceptical because fog had prevented him from seeing any American shore. A second expedition was needed to resolve the issues.

The result was the Great Northern Expedition, intended to determine conclusively the geographical relationship between Asia and America. Bering's company left St Petersburg in 1733, but new port facilities had not been built at Okhotsk, and it was 1740 before Bering's two ships, *Sv Petr* and *Sv Pavel*, were finally launched. Only in 1741 were they able to sail from Petropavlovsk, the new port on Kamchatka. Shortly thereafter, they separated. *Sv Pavel* reached the coast of Alaska, but when captain Alexei Chirikov sent two boats ashore for water, they never returned. Having no more boats, he returned to Petropavlovsk. Bering, meanwhile, also

SOUTH TO JAPAN

The one party of the Great Northern Expedition operating south of Bering's detachment was ordered to investigate the area between Kamchatka and the Kuril Islands. The commander, Martin Spanberg, travelled ahead of Bering to Okhotsk and took charge of the construction of his three ships, which were supposed to have already been built. He then took his small fleet to the Kuril Islands, following which two of the ships continued to Japan, reaching the main island of Honshu.

reached Alaska, where, to the horror of his scientist Georg Steller, he allowed only one day for scientific study before turning back. On the return, the ship was wrecked on what was later named Bering Island. During a terrible winter, 33 men – including Bering – died of exposure and scurvy. The next spring the survivors built a new ship and returned to Petropavlovsk.

THE GREAT NORTHERN EXPEDITION

Bering's detachment was just one of seven parties of the Great Northern Expedition. Five of the parties were ordered to make geographical and scientific surveys of the Siberian coasts north of Bering: Arkhangel'sk to the River Ob', the Ob' to the Yenisey, the Yenisey to the Taymyr peninsula, the Taymyr peninsula to the Lena, and the Lena to the Anadyr' River. Although Bering had nominal overall command, he had little practical authority, and the five groups covering the Arctic coast had widely varying degrees of success in contributions to cartography, hydrography, biogeography and ethnography.

LEFT The Kuril Islands are a rugged chain in the Sea of Okhotsk. In 1738 Martin Spanberg made a reconnaissance of the northern islands, before sailing on to Japan. However, his logs and charting were so poor that the Russian Admiralty ordered him to repeat the voyage so better maps could be produced. In May–June 1742, he sailed south with four ships and, despite problems with fog, extended his discoveries in the Kurils.

ARABIA AND THE MIDDLE EAST

Although Arabia is far closer to southern Europe than are the Americas, the Orient, or much of Africa, there was comparatively little European exploration of it, or of the Persian Gulf region, until the nineteenth century. This was because of the lack of trade opportunities, the hostile terrain and the even greater hostility of many Moslems to Christians. Eventually, most European explorers adopted Arab customs, clothing, language and religion in order to function, or, if they could, went disguised as Islamic pilgrims.

OPPOSITE Harry St John Bridger Philby joined the British foreign service in 1917, and was sent to Arabia where, over the years, he undertook many desert journeys. He was the first European to visit the southern provinces of the Nejd and, on camelback, he mapped what is now the Saudi-Yemeni border in the Rub'al Khali, or "Empty Quarter" (bottom right of this map from 1933), where daytime temperatures of 52°C (126°F) are not unusual. Dissatisfied with British foreign policy in the Middle East, he resigned from the foreign service in 1939, converted to Islam, and took the name of Hajj Abdullah.

LEFT William Gifford Palgrave, who, disguised as a Syrian doctor, became only the second European to explore the northern Nejd, in the heart of Arabia.

MIDDLE In 1806, Johann Burckhardt offered his services to the African Association to penetrate the Sahara from Cairo and make for Timbuktu. His offer was accepted, leading to Burckhardt's eight years in Asia and Africa. However, he died shortly before the caravan he was joining finally departed into the Sahara.

RIGHT Charles Montagu Doughty adamantly refused to change his difficult writing style in *Travels in Arabia Deserta*, but after many rejections it was finally published and became a classic.

One of the first to do this was Lodovico di Varthema. In 1502, he travelled to Cairo, Beirut, Aleppo and Damascus, from where he joined a Mameluke garrison and accompanied a pilgrim caravan to Medina and Mecca, becoming the first European to give accounts of the holy cities. Varthema was arrested in Aden as a Christian spy but eventually made his way to Persia and Afghanistan before continuing an epic journey around India and farther east, finally returning to Europe in 1508.

A century later, two English brothers, Anthony and Robert Sherley, went to Persia to promote trade and form a political alliance. Neither goal came to fruition, but the Sherleys, born adventurers, joined the service of Shah Abbas. In 1600, Anthony crossed the Caspian Sea, sailed up the Volga to Moscow, and returned to Europe via the Muscovy Company route through Arkhangel'sk. Robert remained in Persia and was eventually made an envoy of the Shah.

Although the German surveyor Carsten Niebuhr

RICHARD BURTON
IN DISGUISE

One man fascinated by Burckhardt's
achievements was a British lieutenant in the
Indian Army, Richard Francis Burton. Burton
used his remarkable linguistic talent – he
mastered approximately 40 languages and dialects
– and his understanding of Islamic culture to
visit Medina and Mecca in disguise. In April 1853,
he joined a pilgrim ship from Suez and entered
Medina. He then journeyed to Mecca by an
inland route never previously reported in Europe.
Burton's faultless disguise and behaviour, and
the brilliant writing in his account, gained him a
fame, which further increased when he turned
his attention to Africa.

RIGHT Richard Burton in the disguise
in which he penetrated Medina and
Mecca. He later added the Forbidden
City of Harar to his list of conquests.

ABOVE LEFT Armin Vámbéry
was a remarkable Hungarian
scholar and linguist who
travelled throughout Persia,
Armenia, and Turkestan in the
1860s disguised as a Sunni
dervish. It is thought by some
that he was also a British agent.

ABOVE RIGHT Bertram Thomas
spent several years preparing to
cross the Rub' al Khali, including
making several other remarkable
exploratory journeys, one a 600-
mile camel journey in Dhofar.

spent 1761–67 in Egypt, Arabia, India and the Middle East as part of a scientific expedition sent by Denmark's Frederick V – the last three years as its sole survivor – it was not until the early 1800s that Europeans entered the region in strength. The Spaniard Domingo Badia y Leblich and the German scientist Ulrich Jasper Seetzen each spent years in Arabia and entered Mecca in disguise. But the greatest of their contemporaries was the Swiss traveller Johann Burckhardt, who made an eight-year journey in 1809–17. He visited Damascus and Aleppo and was en route to Cairo when he became the first European in modern times to see Petra. After travelling through Egypt and Nubia, Burckhardt entered Mecca, where he remained for three months, following which he visited Mount Arafat, Medina and Sinai before he died from dysentery in Cairo in 1817.

Burckhardt was followed by several explorers with military backgrounds. In 1819, Captain George Sadleir was sent to contact Ibrahim Pasha, who, having just completed a successful military campaign in Arabia, the British hoped would cooperate in clearing the Persian Gulf of pirates. When Sadleir reached the Gulf, he found Ibrahim had left for Mecca, so he followed him virtually all the way there, becoming the first European to cross the Arabian Peninsula. A decade later, James Wellstead of the East India Company carried out extensive surveys in Oman, Aden and Yemen. And in 1853, Richard Burton penetrated both Mecca and Medina disguised as a Moslem.

Like Burton, many who followed him were enraptured by Arabia. William Gifford Palgrave, an Oxford-educated spy serving both the Pope and Louis Napoleon of France, made the first west–east crossing of Arabia in 1862–63, and became the first European to enter Riyadh. Charles Montagu Doughty spent two years (1876–78) wandering through Arabia, frequently beaten, robbed and starving due to his refusal to conceal either his Christianity or his contempt for Islam. His book *Travels in Arabia Deserta* is a classic of travel writing. And Wilfred and Anne Blunt became life-long campaigners for Arab causes after their travels, which included seeking Arab horses to breed.

PETRA

In August 1812, while travelling through southern Jordan, Johann Burckhardt passed into a narrow gorge and found what the poet William John Burgon later described as "a rose-red city, half as old as time". Burckhardt identified the magnificent rock temples and dwellings carved in the cliff face as the ruins of Petra. The capital of the Nabataeans from the fourth century BC, Petra had been a great trading centre until annexed under the Roman emperor Trajan in AD 106. Burckhardt was the first person from Europe to see it in more than 1,500 years.

LEFT Although a visit to Petra is on numerous travel itineraries today, when Johann Burckhardt went there, the ancient trading centre had disappeared from all but fable in the west.

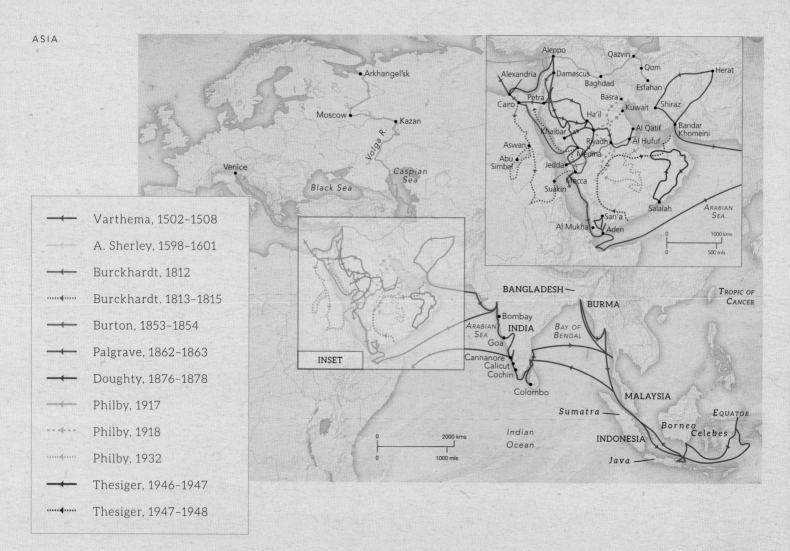

Legend:

- ◄— Varthema, 1502–1508
- ◄— A. Sherley, 1598–1601
- ◄— Burckhardt, 1812
- ·····◄····· Burckhardt, 1813–1815
- ◄— Burton, 1853–1854
- ◄— Palgrave, 1862–1863
- ◄— Doughty, 1876–1878
- — Philby, 1917
- ---◄--- Philby, 1918
- ·····◄····· Philby, 1932
- ◄— Thesiger, 1946–1947
- ·····◄····· Thesiger, 1947–1948

Following the First World War, Harry St John Philby – the father of double-agent Kim Philby – made pioneering journeys on camel. He then planned the first crossing of the desolate Rub' al Khali – the Empty Quarter – the largest continuous sand desert on Earth, then considered the last unexplored land area in the world. To Philby's dismay, the prize was snatched in 1931 by Bertram Thomas, a political officer for the Sultan of Muscat. The next year, Philby finally made extensive explorations. However, the Empty Quarter was not fully investigated until the 1940s, when Wilfrid Thesiger twice crossed it.

RIGHT The main square of the Holy City of Mecca, the home of the Kaaba, the small, black, cubed building towards which Moslems face five times every day in prayer, and which, according to the Qur'an, Adam built as the first house in which to worship Allah.

OPPOSITE This map was drawn by Harry St John Bridger Philby, a British explorer, foreign service official, and author.

Mecca (Makkah in Arabic) is the birthplace of both the Prophet Muhammad and the religion he founded. All adult muslims whose health and finances permit it are required to visit the city at least once in their lifetime, a pilgrimage known as the Hajj. Non-muslims were not permitted in Mecca (this is still the case, with a few exceptions), but Philby's access was legitimate since he had converted to Islam in 1930.

ASIA

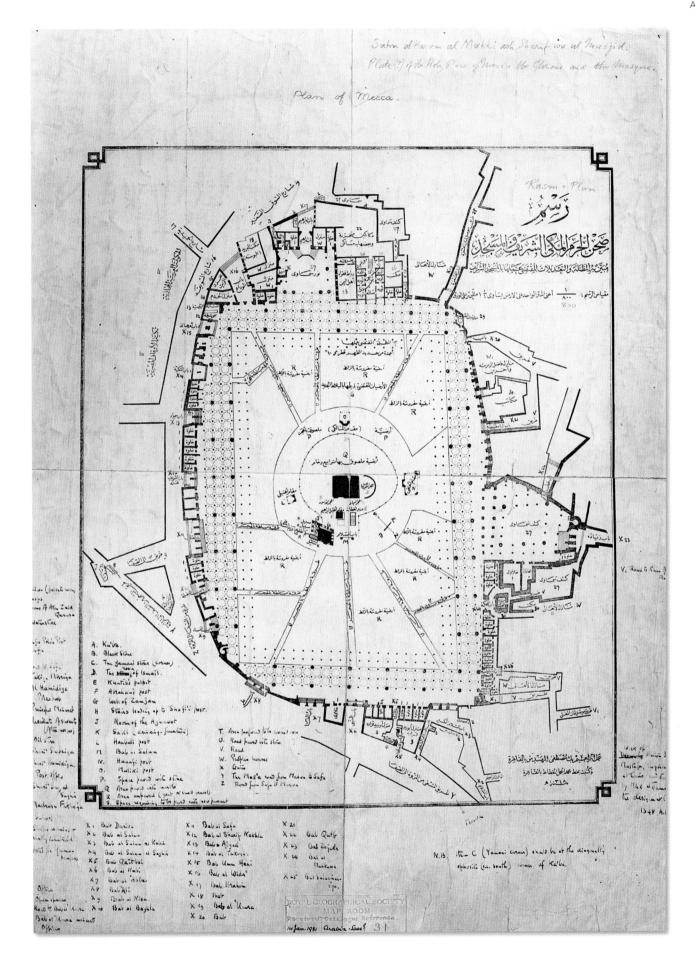

77

LA CONQUÊTE DU THIBET
Entrevue d'officiers anglais avec les Thibétains

THE GREAT GAME

Throughout the nineteenth century, British India and Tsarist Russia eyed each other warily across the ever-decreasing no-man's land between their realms. Each made efforts to explore and annex the intervening regions, and each mistrusted the other's intentions. Geographical exploration became increasingly a cover for espionage, and it became hard to differentiate between government agents and amateur explorers. This led to the activities being called "the Great Game", a term made famous in Rudyard Kipling's *Kim*.

OPPOSITE TOP LEFT An illustration from *Le Petit Journal* in 1904 showed invading troops interacting with the Tibetans after a British-commanded military force stormed into the Hidden Kingdom. The Tibetans initially tried to halt the British advance, but their defensive efforts were quickly overcome by the much greater western firepower.

OPPOSITE TOP RIGHT The first Pundit, Nain Singh, whose unimaginative code name was "Number One". On his first major journey he reached Lhasa and made the first precise determination of its longitude, latitude, and altitude.

OPPOSITE BOTTOM In 1812, William Moorcroft and another adventurer, Hyder Young Hearsey, entered Tibet disguised as Hindu trading pilgrims. There they became the first Europeans to see Manasarovar and Rakas Tal, two sacred lakes of Tibetan Buddhism.

RIGHT Przhevalski's horses, the wild horses of central Asia named for the great Russian explorer who first recorded them. Przhevalski also discovered wild camels, collected zoological and geological samples, and described some 15,000 plants.

One of the earliest episodes in the Great Game came in 1810 when Charles Christie and Henry Pottinger – Indian Army officers disguised as Tatar horse traders – travelled throughout Sindh, Baluchistan and Persia in an effort to gain strategic intelligence that would be of benefit should the Russians advance into the region from the Caucasus.

The Russians' main focus, however, became Central Asia, where they came under the scrutiny of William Moorcroft. The superintendent of the East India Company's horse stud programme, Moorcroft in 1812 became the first Englishman to cross the central Himalayas, while unsuccessfully searching for wild horses to revitalize the bloodlines of the Company's breeds. Seven years later, he began afresh, looking for the legendary Turkmen horses of Central Asia. For six years he sought the horses, while concurrently mapping and recording observations

about Kashmir, the Hindu Kush, Afghanistan and Bukhara. Throughout the period Moorcroft advocated annexing the area before the Russians did, but in 1825 he and his companions died mysteriously.

Moorcroft was followed by a series of daring and resourceful young political officers with a gift for languages and a desire for adventure. Perhaps the most successful of these was the Scotsman Alexander Burnes, who, in 1831, took a barge 1,125 kilometres (700 miles) up the Indus to Lahore, supposedly to deliver presents to a local ruler, but in truth investigating the river's military potential. The next year he made a dangerous trip to Kabul, continued on to the strategic city of Balkh, and then proceeded over the Hindu Kush to Bukhara. His military, political and geographical reports influenced future British policy, and in 1837 he returned to Kabul, where he was murdered in 1841.

Another key figure was Captain T. G. Montgomerie, who initiated the rigorous training of a series of native surveyors. Known as "Pundits," these men could go where Europeans could not. They carried simple, concealed instruments to make observations,

LHASA: THE FORBIDDEN CITY

For many nineteenth-century Europeans, Tibet's mysterious capital was as alluring an objective as Mecca, but Lhasa proved even more elusive. The great Przhevalski almost reached it, but was stopped five-days' ride away. In 1888, unsuccessful attempts were made by the American William Rockhill and an English clergyman, Henry Lansdell, whose letter of introduction from the Archbishop of Canterbury proved unhelpful. Two English women – Annie Taylor and Mrs St George Littledale – and a Canadian, Susie Rijnhart, all headed for Lhasa in the 1890s, but all failed, as did other attempts from at least nine countries.

←— Moorcroft, 1819–1825	←— Przhevalski, 1872–1873
←— Burnes, 1831–1833	·-◄-· Przhevalski, 1876–1877
←— Kishen Singh, 1878–1882	····◄··· Przhevalski, 1879–1880
←— Younghusband, 1887	⁓⁓◄⁓ Przhevalski, 1883–1885

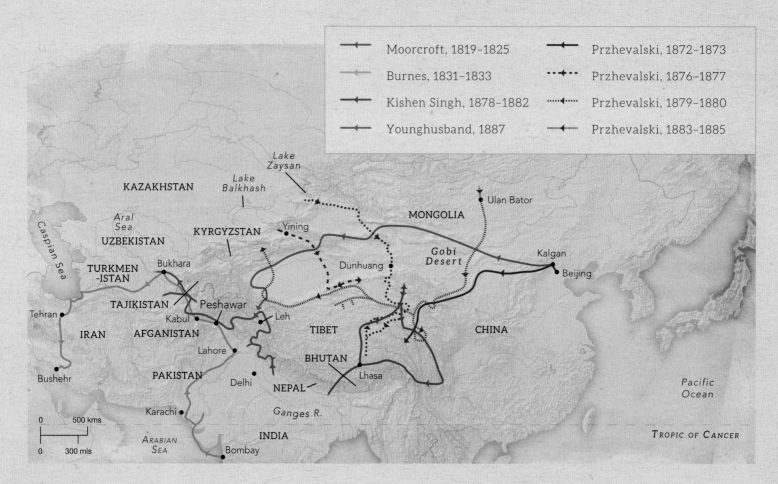

had elaborate methods of hiding their notes and measured distances by recording every step they took on the beads of a Buddhist rosary. One of the first was Nain Singh, who in 1865–66 travelled more than 2,000 kilometres (1,250 miles) through the Himalayas. But the longest journey – more than 4,750 kilometres (3,000 miles) – was made by Kishen Singh. In 1878–82, he crossed the Himalayas to Lhasa, continued through Tibet to Dunhuang on the Silk Road, and returned via western China to find he was officially considered dead.

Meanwhile, the greatest Russian explorer was Nikolai Przhevalski. Setting out from Irkutsk in Siberia in 1870, he traversed Mongolia and the Gobi Desert *en route* to Beijing, from where he turned southwest and crossed China into Tibet, reaching the remote Lake Koko Nor in 1873. Przhevalski led three additional major expeditions, making more comprehensive investigations of Central Asia than anyone previously. He completed the first north–south crossing of the Taklamakan Desert, became the first European since Marco Polo to reach Lake Lop Nor, and made several unsuccessful attempts to reach Lhasa.

The man who finally lived Przhevalski's dream of entering Lhasa was the Englishman Francis Younghusband. He first received notoriety in 1887 with a 1,925-kilometre (1,200-mile) crossing of China, during which he traversed the Gobi Desert and the Tien Shan before making a winter trek over the Karakoram. Younghusband was thereafter a key field operative in the Great Game. In 1903, worried about

OPPOSITE In the late nineteenth century, Lhasa was the most unattainable city on Earth. Today, even its sacred Portola is visited by tourists from around the world.

ABOVE RIGHT Among the diverse discoveries made by Nikolai Przhevalski was that the boundary of the Tibetan Plateau lay approximately 185 miles north of where it had been assumed to be.

RIGHT Francis Young husband's ventures into the unknown regions of Central Asia were generally marked as much by military intelligence gathering and imperial expansion as by exploration.

Russian intervention in the Hidden Kingdom of Tibet, the British invaded it. They quickly overcame Tibetan resistance with modern weaponry, and in August 1904, with Younghusband in charge of political negotiations, military columns marched into Lhasa.

THE GREAT TRIGONOMETRICAL SURVEY

One of the most important aspects of consolidating colonial rule was defining exactly what was ruled. Therefore, in 1800, William Lambton was authorized to start a detailed survey of the southern part of India. By 1805, he had determined that India was actually 65 kilometres (40 miles) narrower than shown on maps, and his success, despite adverse conditions, led to the survey being extended all across India. When Lambton died, he was succeeded by George Everest, who carried the survey into the Himalayas, where his name is immortalized on the highest peak in the world.

LEFT George Everest, the irascible, hard-driven head of the Great Trigonometrical Survey, completed the south–north triangulation of India, thus finishing the longest arc of the meridian yet measured.

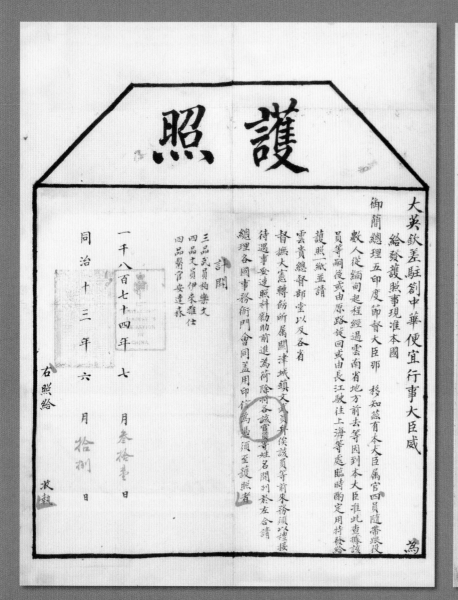

ABOVE The Chinese passport of Ney Elias from the 1880s. Called by Francis Younghusband "the best traveler there has ever been in Central Asia", Elias led eight major expeditions in Central Asia, including crossing sections of China and the Gobi Desert never before seen by Europeans, and being the first Briton to cross the Pamirs. (See Translations, page 205.)

Map Showing the
Route Survey from
NEPAL TO LHASA
and thence through the
UPPER VALLEY OF THE BRAHMAPUTRA
Made by Pundit.
from the Map compiled by Capt.T.G.Montgomerie.R.E.

ABOVE In the second half of the ninteenth century, the British undertook a detailed survey of India. Some of the Indian border countries would not allow Westerners to enter their countries. Thomas G Montgomerie, an officer of the Royal Engineers attached to the survey, overcame this problem by training locals to be surveyors. Called pundits, they were hand-picked for their intelligence and resourcefulness, and trained in clandestine surveying techniques. A pundit travelled the route, surreptitiously noting distances, landmarks and any other significant information. Back at headquarters, the surveys were collated and, slowly, maps of the country were put together. Montgomerie prepared this map in this way and it was published in the *Journal of the Royal Geographical Society* in 1868.

SVEN HEDIN

When Adolf Erik Nordenskiöld returned to Stockholm in 1880 after completing the first navigation of the Northeast Passage, one of those most awed by the celebrations was 15-year-old Sven Hedin, who immediately decided to become an Arctic explorer. He read voraciously about the area, but as a student Hedin's true talents were as a linguist and a mapmaker, both of which would prove key for his life's work.

ABOVE Hedin mounted on a Bactrian camel. Note Hedin's dark glasses; he suffered a serious eye infection in his twenties, and retained limited vision in his left eye, depending solely on his right eye to read and write.

OPPOSITE Sven Hedin's slight frame and bookish appearance belied his physical strength and stamina, exceptional resolve and ambition, and ruthless determination, and led to him be proclaimed "the greatest explorer in the world".

At 20, he was asked to tutor the son of a Swedish engineer in Baku on the Caspian Sea. After doing so for eight months, Hedin made a solo, 3,200-kilometre (2,000-mile) trip on horseback through Persia and Mesopotamia. Afterwards, he found he still wanted to be an explorer, but he now set his sights on Asia rather than the Arctic.

After a preliminary expedition across Russian Central Asia and into the edges of Sinkiang (then known as Chinese Turkestan) in 1890–91, Hedin prepared himself for future scholarly investigations by earning a doctorate in geology and geography. He also published the first of his 50 or more books

about his travels. In late 1893, Hedin began a major expedition, travelling through Russia and then crossing the Pamirs – called the "Roof of the World" – to Kashgar in Sinkiang.

In early 1895, having heard rumours of lost cities in the midst of the terrible Taklamakan Desert, Hedin set out to find them. He had not counted on the ferocity of this no-man's land, however, and his party ran out of water. All but three died, but he had learned many lessons. Later that year, Hedin again entered the Taklamakan, and deep in the desert he found ruins buried in the sand. He excavated several, establishing that a Buddhist culture had existed while

the Silk Road had been a major trade artery through the region, from the fourth to eighth centuries AD. Hedin then ascended the Kunlun Mountains to investigate the northern reaches of Tibet, and continued east to Beijing, from where he travelled north through Siberia.

In 1899, Hedin returned to Sinkiang, where he surveyed and mapped the Yarkand-daria, the river to the north of the Taklamakan. He also investigated the mysterious Lake Lop Nor – which he eventually proved had changed its position through the centuries – and then made his greatest archaeological find, the fabled city of Loulan. Hoping to reach Lhasa, Hedin ascended to the high Tibetan plateau disguised as a Mongolian priest. But only five days from the Holy City, he was stopped by Tibetans. He was ordered to leave the Hidden Kingdom, but Hedin journeyed west instead and then crossed into India.

Even after Francis Younghusband became the first modern European to enter Lhasa, Hedin remained fascinated with Tibet, and in 1906 he overcame the obstinacy of British officialdom and entered it from

TOP During his second expedition to Sinkiang, Hedin surveyed and mapped both the Yarkand-daria, the river running north of the Taklamakan, and Lake Lop Nor. Much of this was done from small, folding boats, such as this.

ABOVE As shown engraved on the case, Hedin lost this tape measure north of Lake Lop Nor in 1901. Five years later, it was found by the archaeologist Aurel Stein, who returned it to Hedin.

RIGHT A map created by Sven Hedin of part of his route in Central Asia on his 1894–97 expedition. Hedin was a talented artist and cartographer, and produced hundreds of maps and drawings during each expedition.

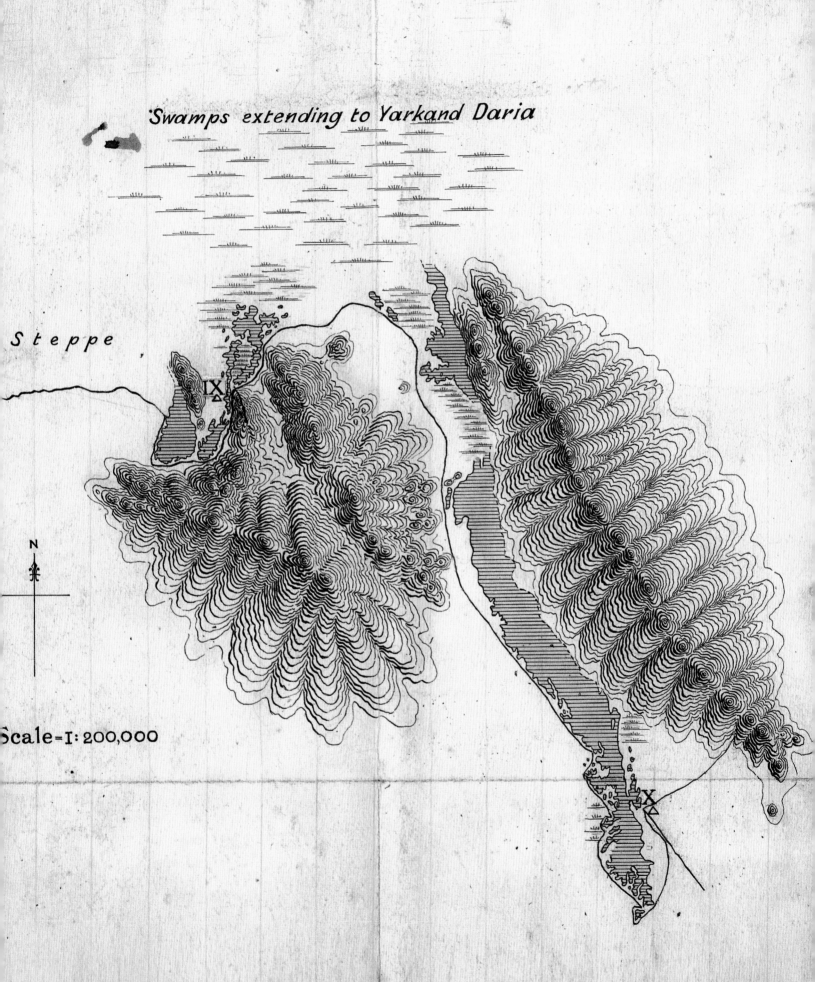

Swamps extending to Yarkand Daria

Steppe

IX
△

N

Scale = 1 : 200,000

X
△

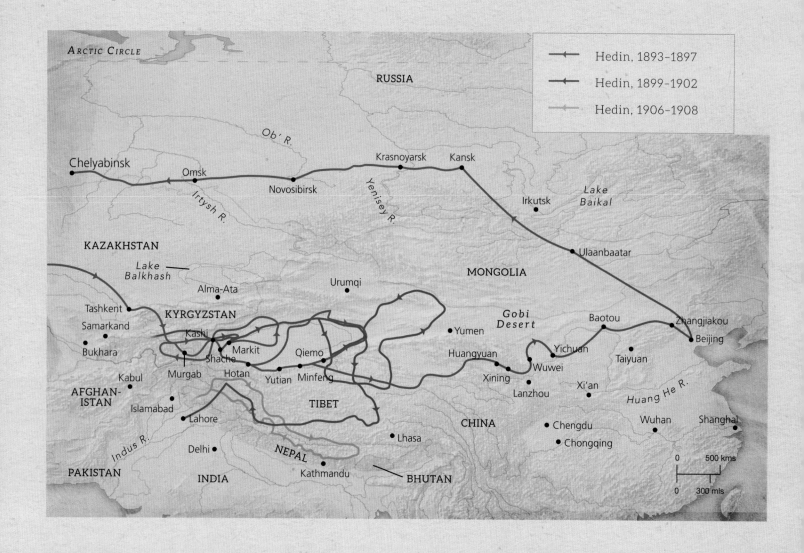

the south. For the next two years, he overcame local attempts to force him to leave the country. Instead, he surveyed and mapped the massive mountain chain north of the Himalayas, which he called the Trans-Himalaya, and the sources of the great Brahmaputra and Indus rivers.

On Hedin's last major effort in Sinkiang, he commanded a large undertaking. The Sino-Swedish Scientific Expedition was designed to lay the groundwork for a German–Chinese air connection, while concurrently carrying out an extensive programme of archaeological, botanical, geological, geodetic and ethnographic studies. However, there was strong Chinese opposition to the expedition

because many European archaeologists had removed ancient treasures to the West. Acting mainly as overseer, diplomat and fund-raiser, Hedin kept the expedition running from 1926 to 1932, and the data obtained eventually filled 54 volumes of scientific results.

For most of his life, Hedin believed that Germany was the paragon of virtue, and he was criticized for supporting Kaiser Wilhelm II in the First World War. Later some considered him a Nazi collaborator, although most of his interactions with German leaders were brought about by his desire to help individuals such as Jewish geographers and Norwegian resistance fighters. He died in 1952.

ABOVE Some of the inspiring scenery at the edge of the Silk Road, where it drops south from Kashgar on its way towards Yarkand and around the southern edge of the Taklamakan Desert.

RIGHT Although there were numerous scientists on the Sino-Swedish Expedition, Hedin's vast experience made him the master in the field. Here he is shown teaching several Mongol members of the expedition how to use a spirit cooker.

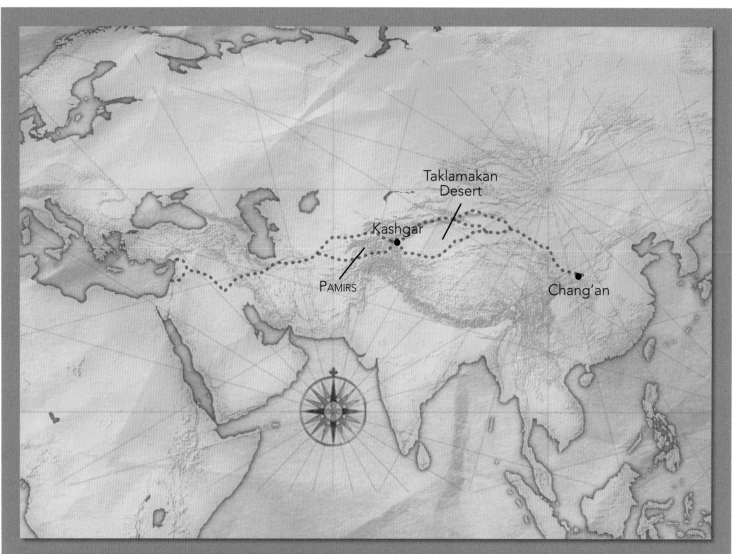

Taklamakan
Desert

Kashgar

PAMIRS

Chang'an

THE SILK ROAD

For thousands of years, trade caravans packed with precious commodities moved between China and civilizations in Central Asia, Persia and further west. Heading west from Xian, the Silk Road – named by geographer Ferdinand von Richthofen – divided into three routes across the Taklamakan Desert, reuniting at Kashgar. It proceeded over the "Roof of the World" and on to the Mediterranean or south into India. The Taklamakan at the time enjoyed a temperate climate and settlements were established on the routes. But over the centuries, progressively warmer climates meant their extinction and disappearance under the great shifting sands, with the last phase of the decline beginning in the 4th century AD.

RIGHT Aurel Stein frequently followed in the footsteps of Sven Hedin – digging at sites Hedin discovered, such as Dandan-Uiliq and Loulan. However, whereas Hedin was an explorer, geographer, and cartographer who engaged in some archaeology, Stein was the opposite: an archaeologist who explored primarily in order to reach archaeological sites.

The tape-measure was found by me on Dec. 23, 1906, lying exposed on the base of the ruined Stupa (relic-tower) rising at the main group of ruins situated in the desert seven marches to the north of Lop-nor marshes and two marches to the south of the salt springs of Altmish Bulak. It had been forgotten there by Dr. Sven Hedin when he first visited the ruins and camped close to the Stupa in March, 1901.

Wind erosion has lowered the ground level to the N and NW of the Stupa by fully 18 feet since the site (now proved to have been a small fortified station on the ancient route from the Tarim to Tun-huang in Kan-su) was deserted in the 3rd century A.D. Being left at the S.W. foot of the massive ruin this recent "archaeological object" was well protected against the erosive action of driving sand. —

M. Aurel Stein

Oxford. Febr. 25, 1909.

RIGHT a letter from Aurel Stein explaining how he found Sven Hedin's tape measure at a site north of Lop Nor more than five years after Hedin lost it.

AUREL STEIN

Aurel Stein (1862-1943) was the giant of Central Asian archaeology. Born in Budapest, he held a series of educational and archaeological posts in India, while leading four major expeditions into Chinese Turkestan. Over a period of 30 years, Stein systematically explored the Silk Road, uncovered its lost Buddhist civilizations, and sent to India and Britain thousands of manuscripts, paintings, textiles and other artefacts. Perhaps his greatest discovery was the "Hidden Library" of the Cave of a Thousand Buddhas near Dunhuang, where he found the earliest known printed book, the *Diamond Sutra*, from AD 868.

AFRICA

A new and most Exact map of AFRICA Described by N. Visscher and don into English Enlarged and Corrected according to I. Blcau and Others With the Habits of ye people & manner of ye Cheife citties, yt haue not before. LONDON by E. White. Printed Collowed and are to be sould John Overton at his hose in Little Brittaine neare the Hospitall. 1668

NORTH AFRICA AND THE SAHARA

For centuries, Europeans – mostly Portuguese and Dutch – sailed the coasts of Africa, engaging in both legitimate commerce and the slave trade. But the interior of the continent remained a mystery. That began to change after 1788 when the Association for Promoting the Discovery of the Interior Parts of Africa was founded in London. Almost immediately, the Association sent out two explorers to reach the Niger River, about which little was known regarding its source, direction of flow, or ultimate destination. One, John Ledyard, died in Cairo in January 1789, which at the time did not seem an unreasonable place to start one's investigations. The other, Simon Lucas, was to cross south through the central Sahara, but he turned back after being warned of the dangers ahead. However, a pattern had been established for exploration via the Sahara.

OPPOSITE This map of the African continent, created by John Overton, was intended to be part of a world atlas. The borders are decorated with costumes, town views and portraits of the native kings, engraved by Philip Holmes.

RIGHT A scene over the rooftops of the once-mysterious Timbuktu. This is the same view that Heinrich Barth would have seen from his lodgings in the city in the 1850s.

A decade later, the African Association engaged Friedrich Hornemann to reach the Niger. But in 1800, after joining a caravan at Murzuk, deep in the Libyan desert, Hornemann disappeared. It would be two decades before it was learned that he had made it to the Niger region, where he apparently died of dysentery. African exploration dried to a trickle during the Napoleonic Wars, but thereafter increased, with the route across the Sahara still considered practical.

In 1817, the British government took up the challenge and sent Joseph Ritchie and George Francis Lyon from Tripoli to discover the course of the Niger. But Lyon returned after Ritchie died in Murzuk, so in 1821, Walter Oudney, Hugh Clapperton and Dixon Denham were dispatched to complete the task. Clapperton, a Scottish Royal Navy lieutenant, and Denham, an English army lieutenant, viewed each other as rivals, and their relations deteriorated to the point where they camped separately and communicated only by letter. They did, however, become the first Europeans to reach Lake Chad, following which Denham explored the local area,

while Clapperton continued west to Sokoto.

Unsatisfied with reports from Denham and Clapperton, in 1825 the British government turned to Alexander Laing to penetrate further into the interior and visit the mysterious city of Timbuktu. Deep in the desert, Laing was attacked by Tuareg nomads and left for dead, but he struggled on alone for 650 kilometres (400 miles) until he reached a place where he could convalesce. He entered Timbuktu in August 1826, but was murdered a month later, just after leaving the city.

In the following years, most exploration of the Niger came from the west rather than the north. However, in 1850, an expedition consisting of James Richardson, Adolf Overweg and Heinrich Barth crossed unknown regions of the Sahara to Lake Chad. First Richardson died, then Overweg, but Barth continued alone, covering more than 16,000 kilometres (10,000 miles). He explored Lake Chad and the upper reaches of the Benue River, before following the Niger to Timbuktu. His remarkable descriptions were published in five volumes and set new standards for scientific observation in African exploration.

French exploration efforts in North Africa

ABOVE George Francis Lyon in a Libyan sandstorm. Travellers throughout the Sahara and Arabia learned to dismount and protect themselves from such destructive winds.

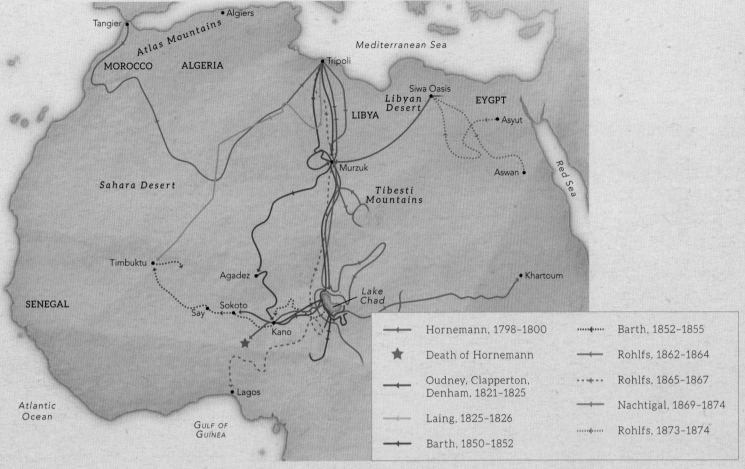

Map legend:
- Hornemann, 1798–1800
- ★ Death of Hornemann
- Oudney, Clapperton, Denham, 1821–1825
- Laing, 1825–1826
- Barth, 1850–1852
- Barth, 1852–1855
- Rohlfs, 1862–1864
- Rohlfs, 1865–1867
- Nachtigal, 1869–1874
- Rohlfs, 1873–1874

increased after that country asserted its claims to much of the region. In 1859, the 19-year-old Henri Duveyrier made the first of several expeditions into the southern Sahara, from which he produced an initial ethnography of the Tuareg. Shortly thereafter, Dr Gerhard Rohlfs, a German in French service, investigated the region south of the Atlas Mountains and unsuccessfully attempted a crossing of the Sahara. He tried again in 1865 and reached Lake Chad, before following the Benue River to the Niger, which in turn he sailed down to the Gulf of Guinea. He later explored the deserts of Libya and Egypt, travelling at times with the German natural historian Georg Schweinfurth.

A third German explorer of North Africa was Gustav Nachtigal, who was appointed by Wilhelm I of Prussia to lead a mission to the Sultan of Bornu in 1869. On his way south, he became the first European to explore the mountains of Tibesti, and after concluding his mission he spent five years investigating the region south of Lake Chad, making forays along the Uele River, and passing through previously unexplored Darfur and Kardofan.

TOP RIGHT Gustav Nachtigal was a physician who moved to North Africa because of a lung ailment. Following his explorations he became president of the Berlin Geographical Society.

BOTTOM RIGHT After his remarkable expeditions to the lands of the Tuareg, Henri Duveyrier served as an influential official with the Geographical Society of Paris.

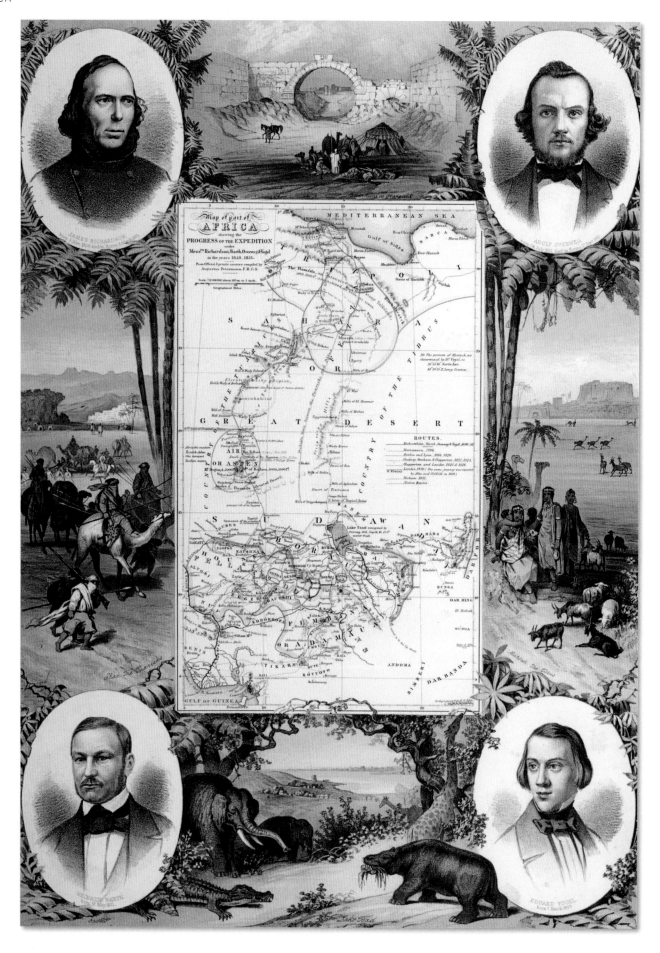

BURCKHARDT IN AFRICA

Johann Burckhardt, the famed Swiss traveller, is best known for discovering the ruined Nabataean city of Petra. However, during an amazing eight-year (1809–17) journey, he also trekked into North Africa. In 1812, Burckhardt reached Cairo, from where he ascended the Nile into Upper Egypt and Nubia. Returning to Aswan, he discovered the partially buried and previously unknown temple of Abu Simbel. Burckhardt then traversed the desert into the Sudan and followed a broad arc east to Suakin, from where he crossed the Red Sea to Jedda in an open boat.

RIGHT The entire complex of the temple of Abu Simbel was moved to higher ground before the Aswan Dam flooded the area.

SIR GEORGE GOLDIE

Although George Goldie's political intrigue and management of the Royal Niger Company in the last two decades of the nineteenth century helped establish the political boundaries of Nigeria, he began his African adventures in the northern deserts. As an extremely rich, wild and dissipated young man, he discarded his army commission, bolted to Cairo and then disappeared into the desert with an Egyptian beauty. In the following years, he learned Arabic, travelled widely and gained an understanding of North Africa's many peoples. He also read Barth's works, which instilled in him the belief that an empire could be built south the Sahara.

LEFT Sir George Goldie – born George Dashwood Goldie Taubman – came from a powerful and long-established family on the Isle of Man.

OPPOSITE The frontispiece to the report by famed armchair geographer August Petermann about the remarkable expedition initially led by James Richardson. Clockwise from top left: Richardson, Adolf Overweg, Eduard Vogel, and Heinrich Barth

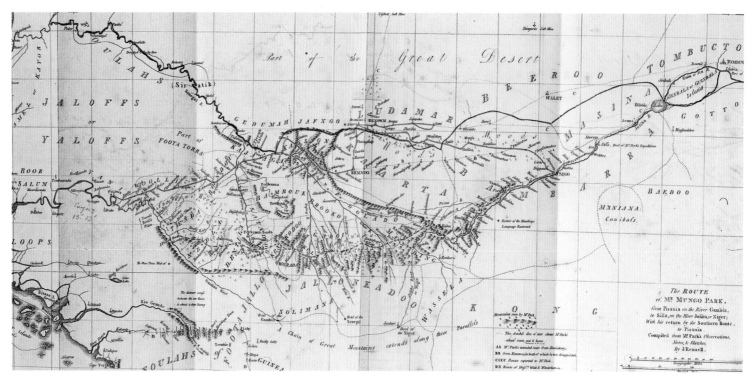

The ROUTE of Mr MUNGO PARK, from Pisania on the River Gambia, to Silla, on the River Joliba, or Niger; With his return by the Southern Route, to Pisania. Compiled from Mr Park's Observations, Notes, & Sketches, By J.Rennell.

MUNGO PARK-LANDER MEMORIAL.

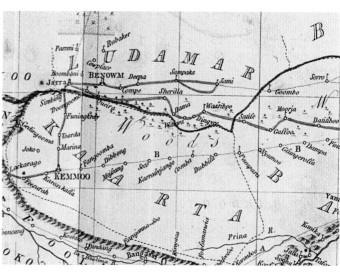

THE WHITE MAN'S GRAVE

Few places in the world posed more difficulties for explorers than West Africa. Its terrain was ferocious, with dense jungles, swamps and waterways that were difficult to navigate. Many native peoples were hostile and the wildlife was a constant threat. But worst of all were the tropical diseases: malaria, blackwater fever, typhus, Guinea worm, dysentery and trypanosomiasis, the last also fatal for pack animals. The region had such a high mortality rate for Europeans that it came to be called the "White Man's Grave".

Nevertheless, explorers continued to try to unveil its mysteries, and nothing was more uncertain than the course of the Niger. Did it link with Lake Chad, the Nile, or the Congo? No one knew. And was the fabled city of Timbuktu truly full of gold? But when the African Association began sending expeditions across the Sahara to find answers, they also dispatched men from the sub-Saharan west coast. The first was Daniel Houghton, who left from the Gambia in 1790; he was subsequently murdered *en route*.

Five years later, the African Association commissioned a Scottish doctor, Mungo Park, to follow Houghton's route and determine the course of the Niger. Leaving from the Gambia, he was robbed several times, then incarcerated by a local Moslem ruler, but he eventually escaped and reached the Niger, which he determined flowed east. He struggled back to the coast despite terrible hardships. A decade later, in 1805, the British government mounted its own expedition, and placed Park in charge. But by the time it reached the Niger, all the pack animals and most of the 45 men had died. Park continued downriver, passing Timbuktu and turning south toward the Gulf of Guinea. Disaster struck at the Bussa Rapids – he and his party disappeared, either killed by hostile locals or drowned.

Another British effort came in 1825. Hugh Clapperton, who earlier had tried to reach the Niger across the Sahara, set out from the Benin coast with his manservant Richard Lander. They crossed the Niger near Bussa and continued to Sokoto, but there, in April 1827, Clapperton died of dysentery. Lander made his way back to England, and his success prompted the government to place him in charge of another expedition – accompanied by his brother John – in 1830. The Landers headed inland from Badagri near Lagos, made their way overland to Bussa, and set sail back down the Niger. After passing its confluence with the Benue, they were taken prisoner by pirates, but were eventually rescued and completed their

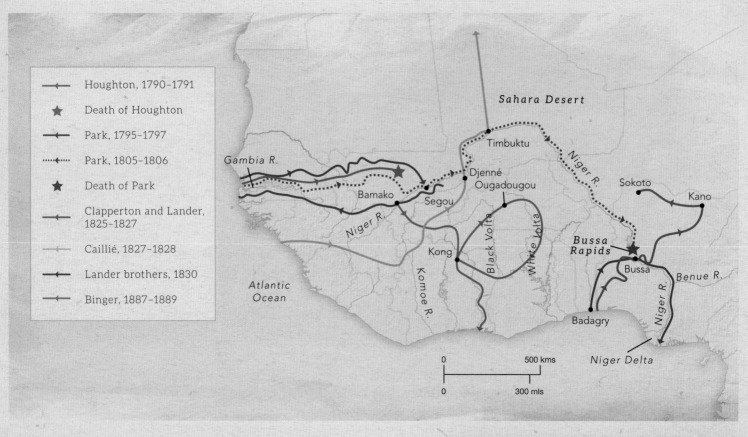

Map legend:

- Houghton, 1790–1791
- ★ Death of Houghton
- Park, 1795–1797
- Park, 1805–1806
- ★ Death of Park
- Clapperton and Lander, 1825–1827
- Caillié, 1827–1828
- Lander brothers, 1830
- Binger, 1887–1889

Labels: Sahara Desert, Gambia R., Timbuktu, Niger R., Sokoto, Kano, Djenné, Ougadougou, Bamako, Segou, Niger R., Bussa Rapids, Bussa, Benue R., Atlantic Ocean, Kong, Black Volta, White Volta, Komoe R., Niger R., Badagry, Niger Delta

0   500 kms
0   300 mls

MARY KINGSLEY

One of the icons of women travellers, Mary Kingsley decided to follow her scientific interests after her parents died within a short time of each other. In 1893–95, she made two extensive journeys to West Africa to study its natural history and native peoples, including their customs, religions and laws. Refusing to "go about Africa in things you would be ashamed of at home", she made difficult journeys in a long skirt, high-necked blouse, and carrying an umbrella. Her acknowledged expertise led to her frequently being consulted by colonial administrators.

RIGHT A famous portrait of Mary Kingsley. She died at 38 from enteric fever, contracted while working as a nurse during the Boer War.

1862 – 1900.

Mary Kingsley.
Scott Wilkinson
CAMBRIDGE.

OPPOSITE RIGHT René-August Caillié, the first European to reach Timbuktu and survive to tell the story. He died in his thirties, probably from disease contracted in Africa.

RIGHT Amber dunes along the Niger River. One of the world's great rivers, its course passes through widely divergent geographical regions occupied by numerous different native peoples.

DU CHAILLU AND THE GORILLA

Much about the early life of Paul Du Chaillu is uncertain, but he definitely lived in Gabon, where his father represented a French firm. After spending several years in New York, Du Chaillu was commissioned in 1855 to report on Gabon's geography, natural history and peoples. On this expedition, he became the first white man to observe gorillas in the wild. Following another expedition, his reports of pygmies were doubted until confirmed by later explorers. Du Chaillu later developed interests in Lapland, and became an expert on Nordic history and culture.

voyage through the delta in early 1831.

Meanwhile, Timbuktu had been reached by a Frenchman. René-August Caillié had attempted to join a British expedition in Senegal at the age of 17, but had been turned down. A decade later, in 1827, he embarked upon a journey to Timbuktu disguised as a Moslem. He finally reached the town to find it had none of the expected glamour or wealth, and he was soon told of Alexander Laing's murder there 18 months before. Two weeks later, Caillié joined a caravan heading for Morocco and returned to France via the Sahara.

The Niger now became a centre of trade rather than exploration, with competing companies setting up bases along the river and its main tributaries. A major figure in this process was Dr William Baikie, who was in charge of a British station at the confluence of the Niger and Benue. In seven years (1857–64), Baikie proved the efficacy of quinine against malaria, using the drug to keep him and his men virtually free of the disease.

But it was the Frenchman Louis-Gustave Binger who did more than any other explorer to open up the country within the great bend of the Niger. In 1887–90, he surveyed the area between the Niger valley and the Black Volta River. Binger thereafter determined the boundary between the British Gold Coast and the French Ivory Coast.

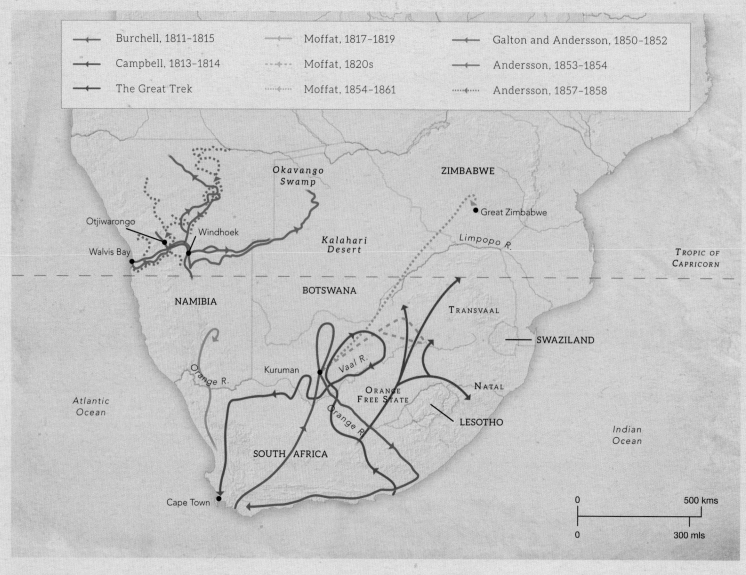

Okavango Swamp

ZIMBABWE

Great Zimbabwe

Kalahari Desert

Limpopo R.

TROPIC OF CAPRICORN

Otjiwarongo

Windhoek

Walvis Bay

NAMIBIA

BOTSWANA

TRANSVAAL

SWAZILAND

Kuruman

Vaal R.

ORANGE FREE STATE

NATAL

Atlantic Ocean

Orange R.

Orange R.

LESOTHO

Indian Ocean

SOUTH AFRICA

Cape Town

0   500 kms

0   300 mls

SOUTHERN AFRICA: A DIFFERENT PATTERN

The unveiling of Southern Africa followed a different process to that of the rest of the continent. Rather than being opened up by explorers, most of it was first penetrated by pastoralists, naturalists, or missionaries. In 1652, the Dutch East India Company established a settlement by Table Bay to serve as a victualling station for its ships. In the following 150 years, the colony grew, and the Boers – settlers of Dutch, German, or French Huguenot descent – moved slowly inland, either pushing out or enslaving the native populations.

OPPOSITE FAR LEFT John Campbell preached to many African peoples but unfortunately found that the number watching him "taking dinner was at least three times greater than attended the worship".

OPPOSITE LEFT John Campbell, as pictured by fellow minister W. T. Strutt. Many of the policies followed by later missionaries were based on Campbell's experiences.

ABOVE The shadow stripe between the main stripes is one way to differentiate Burchell's zebras, as well as that the main stripes tend to be dark brown rather than black.

In the late eighteenth century, a new type of European began to arrive in Southern Africa: men who wanted to find and catalogue new specimens of rocks, plants, animals, birds and insects. Among the earliest of these, in the 1770s, were the Swedish naturalists Carl Thunberg and Anders Sparrman and the Scottish collector William Paterson. Their combined work was in part why the African Association concentrated so heavily on North Africa – the south was considered already explored.

In 1795, the British seized the Cape from the Dutch to prevent any French attempt at occupation, and following the Napoleonic Wars it was decided the Cape Colony would remain in British hands. In the interim, British naturalists and missionaries had made their way farther into the interior. William Burchell spent 1811–15 wandering throughout the

area, and brought back to England samples of 80 different species of mammal, 265 birds and a vast number of botanical specimens, as well as more than 500 remarkable drawings. Meanwhile, the Reverend John Campbell made two lengthy expeditions into unknown territories, crossing the Orange and Vaal rivers and reaching Bechuanaland (Botswana). On these journeys, he worked not only to convert the African population to Christianity, but contributed significantly to knowledge of the region's natural history and geography.

The greatest of the British missionaries in Southern Africa, however, was Robert Moffat, who was sent by the London Missionary Society. In 1817, Moffat travelled north through Namaqualand into what is today Namibia, but, convinced there was no hope of establishing a missionary station there, he returned to

Cape Town the next year. He then was assigned to the Lattakoo mission of the Kalahari Desert; several years later he and his wife moved to Kuruman, even farther into the Kalahari. In the next four decades, Moffat journeyed into Bechuanaland, passed through the Transvaal to the Limpopo River, and even went as far as modern Zimbabwe. Perhaps his greatest contribution was translating the Bible into the Tswana language, but he is remembered more today for his encouragement of the young missionary David Livingstone, who eventually married Moffat's daughter Mary.

The next stage of expansion was again accomplished by pastoralists. The Boers disliked British rule, particularly the social and economic policies, and their dissatisfaction came to a head in 1834 with the British abolition of slavery. Shortly thereafter, between 1834 and 1840, approximately 15,000 Cape Boers made a series of migrations inland, hoping to break away from the British and establish their own republics. These areas later became known as Natal, the Orange Free State and the Transvaal. This mass exodus, called the "Great Trek", did not have exploration as an aim, but it did open up knowledge of vast new regions, which were seized by the Boers while simultaneously displacing the native tribes farther north or east.

In 1850, an expedition was launched from Walvis Bay in Namibia the true purpose of which was geographical and scientific exploration. Francis Galton, an English gentleman-explorer, hoped to cross the Kalahari

and reach Lake Ngami with his assistant Charles Andersson. However, they were turned back by local troubles, so instead they headed north, becoming the first Europeans to investigate Damaraland, Etosha Pan and Ovamboland. Galton then returned to England, but Andersson stayed in southwest Africa. In 1853–54, he finally reached Lake Ngami, and several years later he discovered the Okavango Delta of Bechuanaland while unsuccessfully trying to cross into Angola.

ABOVE The uneasy relationship between the British and Boers exploded several times, with wars fought in 1880–1881 and again in 1899–1902.

FRANCIS GALTON IN ENGLAND

Although overshadowed historically by his cousin Charles Darwin, Francis Galton (1822–1911) was a true Victorian polymath. He had a life-long interest in exploration, and served almost four decades on the Council of the RGS, while his classic book *The Art of Travel* went through numerous editions. Galton is best known for his studies of human heredity and intelligence, innovative work that established the field he named "eugenics". His other contributions included the discovery of the anticyclone, pioneering the systematic comparison of fingerprints and effectively inventing regression and correlation analysis in statistics.

LEFT Of Francis Galton's myriad scholarly contributions, perhaps the greatest came in studying human heredity. His research included innovative methods such as examinations of twins, pedigree analysis, and anthropometric studies. He is also remembered for his bitter feud with the explorer H. M. Stanley.

THE CAPE OF GOOD HOPE

The Dutch first established an outpost at the Cape of Good Hope in 1652. Several years later, Dutch watercolourist and cartographer Johannes Vingboons created this watercolour drawing of Table Bay and Robben Island and included it in his *Manuscript Atlas*. Vingboons used the drawings that captains and crew of the Dutch East India Company took home from their voyages to created hundreds of watercolours from his Amsterdam workshop.

ABOVE Dutch East India Company ships in Table Bay. The Dutch began Robben Island's long history as a prison at this time.

RIGHT William Burchell's collections from Southern Africa were amongst the most impressive ever compiled. He later spent four years in Brazil, but his contributions from there were nowhere near as significant.

THE PORTUGUESE IN THE INTERIOR

In Britain, one of David Livingstone's perceived triumphs was reaching the mighty Zambezi River. However, although their deeds were little known, Portuguese traders and slavers had already explored much of the region from Mozambique or Angola. In 1831–32, José Correia Monteiro followed the Zambezi and then penetrated north to Lake Bangweolu and Lake Mweru. In 1846, the judge Candido de Costa Cardoso made a thorough investigation of Lake Nyasa, which Livingstone later struggled to reach. And shortly thereafter, Antonio da Silva Porto reached the Zambezi near the very place Livingstone would later "discover".lva Porto reached the Zambezi near the very place Livingstone would later "discover".

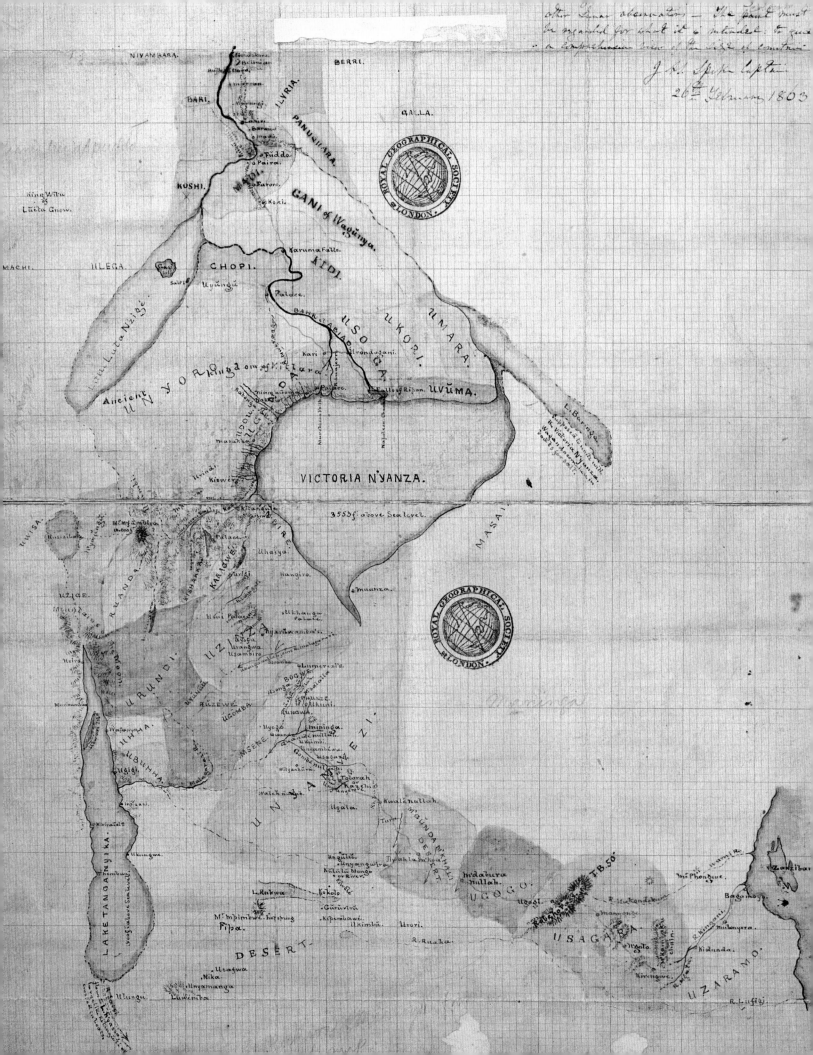

NIYAMBARA.

BERRI.

ILYRIA.

BARI.

GALLA.

KOSHI.

King Witu
25
Lütta Gnow.

MACHI.

ILEGA.

CHOPI.

GANI of Waganya.

Karuma Falls.

Uyungu.

Palace.

BAHR el ABIAD.

USOGA.

UKORI.

UMARA.

Kari.

Urondagani.

Ancient UNYORO kingdom of Kittara.

Falls of Ripon.

UVUMA.

L. Baringa.

UDDU

Masakka.

Kiowey.

VICTORIA N'YANZA.

MASAI.

3553f. above Sea level.

LUISA.

M: Mfumbiro
10.000

RWANDA.

KARAGWE.

Uhaiya.

Hangiro.

UZIGE.

Palace.

Muanza.

URUNDI.

UZINZA.

Usambiro.

Seronia.

BOGWE.

RUZEWE.

UGOMBA.

MSENE.

UNAMUEZI.

Uyogo.

Miminga.

USIGI.

Tabora.

Kazeh.

Waten najib.

Kwalé nullah.

Ugala.

MGUNDA MKHALI DESERT.

LAKE TANGANYIKA.

Tiyuhla n'kari.

M'dabura nullah.

UGOGO.

T.B.50.

R. Wami R.

Zanzibar.

M'Phondwe.

Bagamoyo.

L. Rukwa.

Kokolo.

M: mpimbwe.
Fipa.

Gururu.

Kipembawe.

Ukimbu.

Urori.

R. Ruaha.

USAGARA.

Kidunda.

UZARAMO.

DESERT.

Usagwa.

Nika.

Unyamanga.

Luwimba.

Ulungu.

R. Lufiji.

THE SOURCES OF THE NILE

For thousands of years, people had wondered where the Nile began. The world's longest river, stretching more than 6,600 kilometres (4,100 miles), it had been traced through the deserts, but violent cataracts, malaria and the Sudd – a vast swamp of papyrus reeds and rotting vegetation – had prevented anyone from following it to its source. And so myth and legend persisted: that it was born in a series of inland lakes or, according to the Greek geographer Ptolemy, in the Mountains of the Moon.

OPPOSITE A map produced by James Grant. He and John Hanning Speke did not trace the Nile for its full course, however, so Speke's theories remained unproved.

RIGHT John Hanning Speke (top) and Richard Francis Burton (bottom) after they had become bitter rivals. After Speke's death, Burton continued to travel widely and wrote numerous books about his adventures. However, his journeys never again captured the public's imagination as had his earlier ones to Mecca, Harar, or Lake Tanganyika.

In 1770, James Bruce reached the sources of the Blue Nile. But even by the mid-nineteenth century, little was known about the great southern branch of the river, the White Nile. Then, in 1848, Johann Rebmann of the Church Missionary Society became the first European to see Mount Kilimanjaro, and the next year Johann Krapf, another missionary, spied Mount Kenya. Their reports of snow-covered mountains on the Equator were initially greeted with scepticism, but soon fired a new era of African exploration.

In 1857, Richard Francis Burton, already famous for his remarkable journey to Mecca and Medina, and for an equally perilous one in 1854 to the forbidden city of Harar (in modern Ethiopia), left Zanzibar in search of a vast lake rumoured to be far inland. Joining him was an Indian Army officer, John Hanning Speke. The two men had little in common: Burton had a remarkable intellect and combined broad-ranging scholarship with eccentric, libertine behaviour; Speke was an obsessive big-game hunter. For seven months they struggled west, suffering from malaria and other tropical diseases, before, in February 1858, discovering the amazingly long Lake Tanganyika. Unfortunately, they were too ill to determine

AFRICA

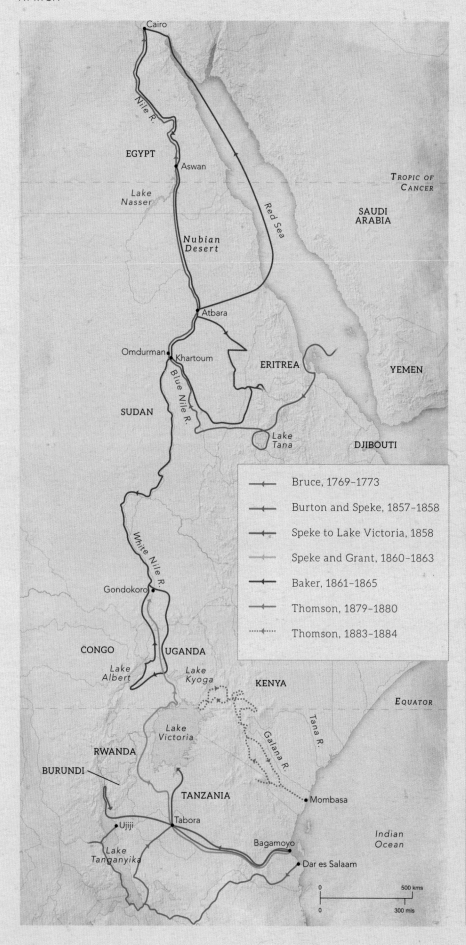

Legend:
- Bruce, 1769–1773
- Burton and Speke, 1857–1858
- Speke to Lake Victoria, 1858
- Speke and Grant, 1860–1863
- Baker, 1861–1865
- Thomson, 1879–1880
- Thomson, 1883–1884

THE BLUE NILE

In 1768, the Scotsman James Bruce set forth to discover the true source of the Nile, which, based on reports by the Jesuit Pedro Paez, he believed originated with the Blue Nile in Abyssinia. After countless adventures, in November 1770, Bruce reached the Tisisat Falls and Lake Tana (previously attained by Paez in 1618) and then ascended the Abay River to the Blue Nile's primary source. Throughout his life, Bruce maintained he had discovered the source of the Nile, although on his return home via Khartoum he found the White Nile was a much greater river. However, in Britain his account was considered too fabulous to be true, and for decades it was not believed.

ABOVE James "Abyssinian" Bruce in old age. Embittered by the many people who did not believe his amazing – but true – tales, he retired to his estate in Scotland.

110

ABOVE No other geographical question perplexed mankind longer than that of the sources of the Nile. For thousands of years, its northern course and delta were intimately known, but its upper reaches remained shrouded in mystery. Even today, few people have experienced its long journey from Lake Victoria to the Mediterranean.

BELOW Speke and Grant addressing the members of the Royal Geographical Society after their triumphant return to London in 1863.

whether the river at its north end flowed into or out of the lake.

They headed back towards the coast, and while Burton recovered at an Arab trading centre, Speke made an excursion north. He returned announcing he had discovered a lake that was the source of the Nile. And so began an historic quarrel, Speke arguing – without foundation, but accurately – that Lake Victoria was the birthplace of the Nile, and Burton countering that it was Lake Tanganyika. After convalescing, Burton returned to England to find that Speke had already presented his version to the Royal Geographical Society.

To Burton's chagrin, the RGS favoured Speke's conclusions, and in 1860 Speke was sent back with James Grant for further exploration. Over the next three years, the two made many discoveries, including Ripon Falls, where the Nile flows from Lake

Victoria. However, they failed to prove Speke's theory because they did not trace the continuous course of the river. The conflicting claims of Burton and Speke resulted in a debate being scheduled at the RGS for September 1864. On the day before the meeting, however, Speke was killed in a shooting accident.

Meanwhile, another piece was added to the Nile puzzle by Samuel and Florence Baker. The couple came to Africa on their own search for the Nile source in 1861, and made their way upriver from Cairo; *en route* they investigated the Blue Nile's tributaries. To the west of Speke's route, they found what Baker named Lake Albert. At its northern end, the Nile flowed into, and then back out of, the lake. Following the Nile upstream, they discovered another great falls, which Baker named after RGS President Roderick Murchison. With the Bakers' discoveries, most of the Nile's mysteries had been solved. However, key sections remained unexplored, and Speke had not proved whether Lake Victoria was an inland sea or a collection of lakes. To settle these issues, in 1866 the RGS sent David Livingstone back to Africa.

Despite Livingstone's expeditions and many others from the coast, much of East Africa remained unknown. Determined to change that, in 1878, the RGS dispatched an expedition to Lake Nyasa and Lake Tanganyika. The leader died, and command was assumed by a 20-year-old Scotsman, Joseph Thomson, one of the truly gentle personalities of African exploration. But Thomson was also one of the most effective, and he subsequently pioneered a new route to Lake Victoria through the land of the Masai.

FLORENCE BAKER

Samuel Baker once declared that he owed "my success and my life" to his wife Florence, and she certainly could have said the same about her husband. In 1859, Baker was travelling through the Balkans, then part of the Ottoman empire. At a Turkish slave market in Bulgaria, he saw a slight, beautiful, 18-year-old Hungarian girl for sale, and on impulse he bought her. They soon married and set out for Africa, where Florence proved the ideal complement to her husband. Her charm, persuasiveness, kindliness and foresight more than once saved them from precarious situations.

BELOW English explorer Samuel White Baker set out in March 1861 accompanied by the woman who later became his wife, on his first Central African exploration. The trip was undertaken, in his own words, "to discover the sources of the river Nile, with the hope of meeting the East African expedition under Captains Speke and Grant somewhere about the Victoria Lake". He spent a year on the Sudan–Abyssinian border exploring the Atbara River and other Nile tributaries (he proved that the Nile sediment came from Abyssinia). He did meet Speke and Grant early in 1863 and, using information they gave him, became the first European to discover Lake Albert in 1864. His book *The Nile Tributaries of Abyssinia* proved very popular.

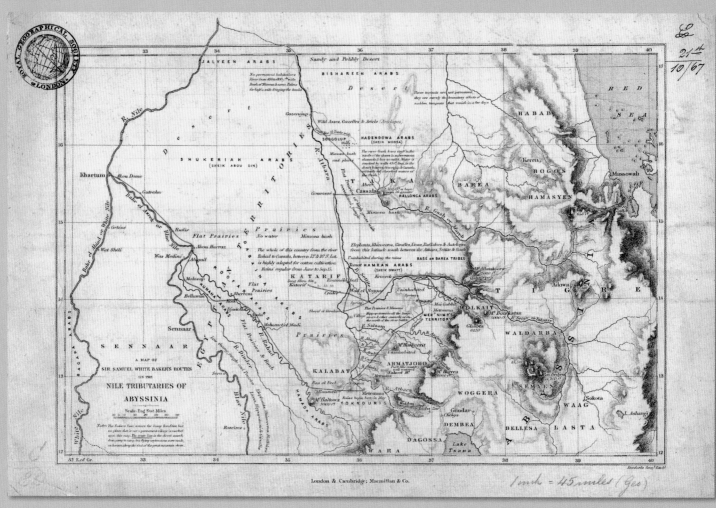

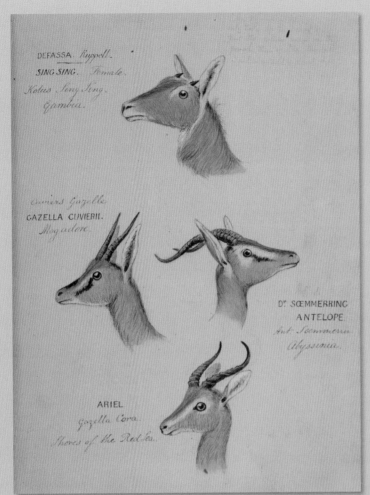

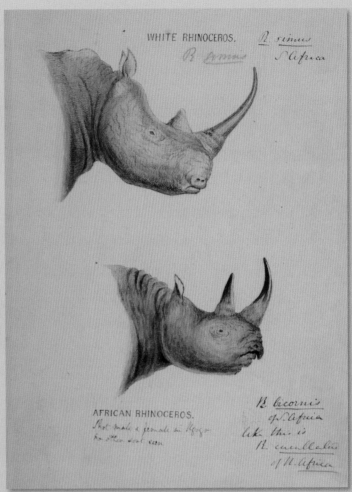

OPPOSITE RIGHT Florence Baker could be every bit as tough as her husband. Once, when a chief offered to swap wives with Baker, she terminated the proposition with "a countenance as amiable as Medusa".

OPPOSITE LEFT Samuel Baker in the hunting suit made by Florence. Baker later became Governor-General of Equatoria, where he tried to eradicate the slave trade.

ABOVE John Hanning Speke was an avid, almost obsessive, hunter but, as these remarkably accurate drawings from his sketchbook show, he was also a talented wildlife artist.

RIGHT Throughout his life, Samuel Baker was an avid hunter, and this was one of his guns, capable of bringing down elephants.

in black pencil are correct in Latitude only

those in black ink are put in on the authority
of my own observations. the Portuguese
spelling is retained except where it would mislead
an Englishman, as for instance Chinge. pro^d
soft, Tinje. Chioque. pno^d Kioke.

Mais chief of a very large an
populous territory. the people
named Maias. (N^m ea ma^
unknown to the Portuguese)

R. Congo or Zaire

Muanzanza
(a sovereign)

Holo- very populous country
the Holo

Mataba (country of

Pende

Pungo Andongo, Port^s fort
Lat. 9.42.28 S.   2 { 9. 40. 8 S
1 { Long. 15. 23. 5 E.   15. 23. 40 E
1 Communicated in Cons. Branch letter, May 18. 1855.
2.   On Livingstons letter Dec 21 1855
        9. 42. 28
        15. 23. 5 E

Memoranda of Latitudes from observations
R. Bengo 20 miles from Loanda 8° 48' 43" South
  Golungo Alto — — — 9° 8' 30" —
  Ambaca — — — — 9° 16' 35" —
  Mohwira he near Litele — 9° 26' 25" —
  Gio or Geo — — — 9° 39' 14" —
  Sanza or R. Luize — — 9° 37' 46" —
  Cassange — — — 9° 37' 30" South Long. 17.45 E.

14. 59. 27 mean of 7 Lunars at residence of

the Holo country reported to be
very populous.

Luba 25 days N.E. of Cassange
each day being between 10 & 12 Geog. miles

the Yungo (Hungo Port^sp) Mountainous country.

Capenda Camubimbe

the Tinje
(Hionge Port^l sp^l)

The Tinje

Tala Canganza   m^tns

The Tinje (Chinge. Portuguese spelling)

Matiamvo
×

R. Dandé

Jinje

Bengo R.

Henge m^ts

The Kioke

Lunda or country of Balo
of whom Matiamvo is par^

is reported by intelligent
to form the Zaire or Congo
flows into it according to

R. Coanza

Cassange

the Cheeboque

Lake
Kipembo

R. Luango
source unknown.
probably the Luare

Plains of Lobale
reported impassable
during the rainy season

Lunda or
of the Balon

Benguela

Lobale

Return Journey
R. Luango   { 9. 49   18. 25
between   { 9. 52   18. 30
R. Chikapa   10. 10   19. 42
R. Maomba   9. 38   20. 13. 30
Cobango   9. 31   20. 31

15 miles East of Braun Ob^d made in
cloudy weather
the course of this river altered by
Obs^t on return journey

Matiamvo is E.N.E. 100 miles.

The Ganguelas

R. Quando R. Coti (Kotee)

R. Leeba

Confluence of the
R. Leeba & Leeambye
Lat. 14° 11' South
Long. 23. 60' 30" East.

R. Looti

Main branch of the
or Zambes

R. Chanabara

R. Quito

R. Coquema
(Kokema)

Caconda

Borotse valley known in
North as L
Milui & M

country of the
Cassagure (Kassakere) the Ganguelas
or real Bushmen

Cunene

R. Chobe

DAVID LIVINGSTON

David Livingstone was the supreme Victorian example of self-improvement. The son of a poor mill-worker in Blantyre, south of Glasgow, at the age of 10 he started working in a mill himself – 13 hours a day, six days a week. Nevertheless, he attended school each night, and during the next decade taught himself Latin, natural history and theology. At 21, Livingstone decided to be a medical missionary, so he saved to attend medical school and thereafter applied to receive training from the London Missionary Society. In November 1840, he received his certification as a doctor and was ordained a minister, following which he left for southern Africa.

OPPOSITE African missionary and explorer, Dr David Livingstone, was one of the first Westerners to undertake journeys across the continent of Africa with the aim of opening the routes, accumulating useful information and establishing Christian mission stations. This map, drawn by Livingstone himself, shows the route of his 1853–4 journey, undertaken in canoes and on ox-back, from the heart of the continent in Sesheke, where the Zambezi river rises, to the Portuguese settlement of Loanda on the west coast.

ABOVE Livingstone as he appeared in 1857 during his triumphant return to Britain. He received the Gold Medal of the Royal Geographical Society, had a private audience with Queen Victoria and his book *Missionary Travels and Researches in South Africa* became a best-seller.

There Livingstone married the daughter of Robert Moffatt, the renowned missionary and explorer. But Livingstone was unsuccessful in converting natives to Christianity and in 1849 headed north with a wealthy traveller, William Cotton Oswell, while reassessing his missionary role. The two men crossed the Kalahari to Lake Ngami, which had never been seen by Europeans before. Having heard rumours of a great river beyond the lake, in 1851 they went north again, reaching the banks of the Zambezi River and discovering that Portuguese slavers had penetrated farther into the interior than anyone in Britain had suspected.

Livingstone now developed a theory about how the mighty Zambezi might solve the horrors of slaving: missionaries and commercial traders could follow it, he thought, from the coast to the interior, bringing with them a multitude of British products to exchange for wax, palm oil and ivory. Such easy availability would mean that the local chiefs would no longer be enticed to trade slaves for European commodities. So he decided to help open up the Zambezi for "legitimate" commerce and missionary work. With this in mind, between 1853 and 1856 Livingstone made the first known crossing of Africa. After again reaching the mid-Zambezi, he traced it west to the coast of

Angola. Then, deciding this mountainous route was too difficult, he followed its general course to the east coast, becoming the first European to reach the great waterfalls of Mosi-oa-tunya, or "the smoke that thunders", which he named Victoria Falls.

When Livingstone returned to Britain in 1856, he found himself a national hero for what Roderick Murchison of the Royal Geographical Society called "the greatest triumph in geographical research effected in our times". With the sponsorship of the RGS, Livingstone was sent back to navigate the Zambezi fully and establish a series of missions. When it proved impossible to reach Victoria Falls by boat, Livingstone changed his plan and spent the bulk of the expedition (1858–64) making an attempt to reach Lake Nyasa along the Shire River, which failed miserably, while an effort to found a mission ended in disaster.

Despite this lack of success, Livingstone remained an enormously popular hero, and the Royal Geographical Society sponsored his return to Africa to solve once and for all the mysteries surrounding

LEFT AND BELOW The hats that were worn by Livingstone and Henry Morton Stanley became part of each man's icon. Livingstone's cap, with the red band, was included in innumerable portraits of the explorer from the time of his first fame until his death. Stanley's early self-publicity frequently included the wearing of his carefully adorned pith helmet.

BOTTOM In 1850, the year after having discovered Lake Ngami, Livingstone returned with his family. He accomplished little, however, and it was not until the following year that he and William Cotton Oswell went further north, reaching the mighty Zambezi River.

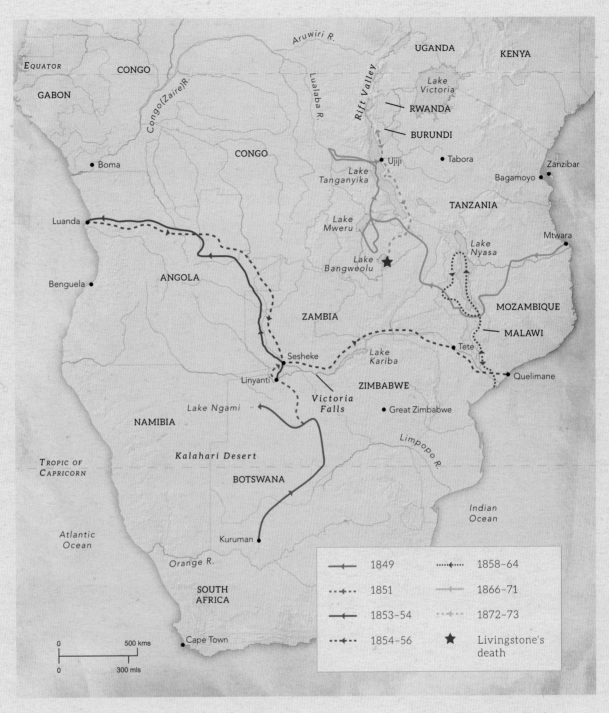

EQUATOR

CONGO

GABON

Aruwiri R.

UGANDA   KENYA

Lake
Victoria

RWANDA

BURUNDI

Congo (Zaire) R.

Lualaba R.

Rift Valley

CONGO

• Boma

Ujiji •   • Tabora   Zanzibar

Lake
Tanganyika   Bagamoyo •

TANZANIA

Luanda

Lake
Mweru   Lake
Nyasa   Mtwara

ANGOLA   Lake
Bangweolu

Benguela •

ZAMBIA   MOZAMBIQUE

MALAWI

Sesheke   Lake
Kariba   Tete

Linyanti   Quelimane

Victoria
Falls   ZIMBABWE

Lake Ngami   • Great Zimbabwe

NAMIBIA

Limpopo R.

TROPIC OF
CAPRICORN

Kalahari Desert   Indian
Ocean

BOTSWANA

Atlantic
Ocean

Kuruman

Orange R.

SOUTH
AFRICA

Cape Town

	1849		1858–64
	1851		1866–71
	1853–54		1872–73
	1854–56	★	Livingstone's death

0   500 kms

0   300 mls

the sources of the Nile. These had been brought to public attention by the conflicting claims of Richard Burton and John Hanning Speke. The next year Livingstone disappeared into central Africa on another expedition, but his name was kept before the public by the mystery of his whereabouts, by his powerful supporters and by his unknowingly being used as the spokesman of the missionary movement, the anti-slavery movement and by expansionists desiring to open up Africa.

In the last year of his life, Livingstone received huge publicity due to Henry Morton Stanley's "finding" him. Despite crumbling health, he refused to return to Britain, and continued his search for the sources of the Nile south of Lake Tanganyika, far from its real beginnings. He died near Lake Bangweolu on 1 May 1873, widely reported to have been seen that night reading the Bible and then found kneeling by his bedside, his face buried in his hands, as if in prayer. The iconic nature of his death helped shape his legend and establish his mythic status as one perhaps without peer among the heroes of the British empire.

Pls send no cop-
ies until further
news from us-

-Unyanyembe-
October 16th 1873

Dear Sir

It is with extreme regret
I write to announce to you the melan-
choly news of the death of Dr Livingstone
of which we received news from Chumoi
his servant, who came in in advance of
his caravan in order to get some as he
says they are ~~i~~ utterly destitute-
                    (Chumois)
From his report, they had passed
the northern shores of Lake Bemba
and arrived at about 10° S Lat on the
Luapula, when the Doctor was attacked
with Dysentery, which carried him off
in about ten days or a fortnight.
His servants have disembowelled
the corpse and filled it with salt
and put brandy into the mouth &c
so as to preserve it and are being-

LEFT A letter, dated 16 October 1873, from Verney Lovett Cameron in Unyanyembe to the Secretary of the Royal Geographic Society in London, announcing the death of Livingstone.

BELOW Livingstone's hypsometrical apparatus, which measures the temperature at which water boils thus determining height above sea level, was also used by Verney Lovett Cameron, the first European to cross central Africa from east to west.

OPPOSITE BOTTOM Livingstone's own watercolour sketch of Victoria Falls, with his accompanying notes explaining what he saw.

BRINGING LIVINGSTONE TO WESTMINSTER ABBEY

There are few greater historical examples of devotion than that of Livingstone's two African assistants, Chuma and Susi. Deciding that the explorer's body should be returned to Britain, they buried his heart and internal organs in the Africa he loved, dried out and preserved his body, and then carried it back on a five-month journey to the coast. This was done despite the knowledge that if they were discovered transporting a corpse, they might be killed as witches. But Livingstone's body reached Britain, and he was buried at Westminster Abbey in 1874.

"DR LIVINGSTONE, I PRESUME?"

In the years after Livingstone vanished into central Africa, numerous expeditions unsuccessfully sought to find him. In 1869, James Gordon Bennett Jr, the proprietor of *The New York Herald*, the largest newspaper in the United States, surmised that finding Livingstone would be a huge journalistic scoop. He therefore sent one of his reporters, Henry Morton Stanley, to locate the explorer, telling him to spend whatever money was required. In November 1871, Stanley reached Livingstone at Ujiji on the shores of Lake Tanganyika, where he uttered his immortal comment.

LEFT In perhaps the most famous encounter in the history of exploration, newspaper correspondent Henry Morton Stanley greeted Livingstone with the question, "Dr Livingstone, I presume?"

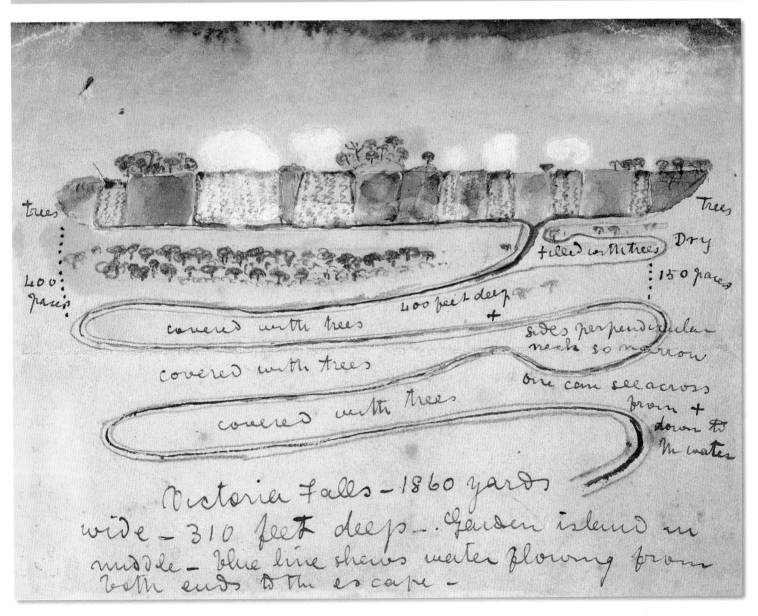

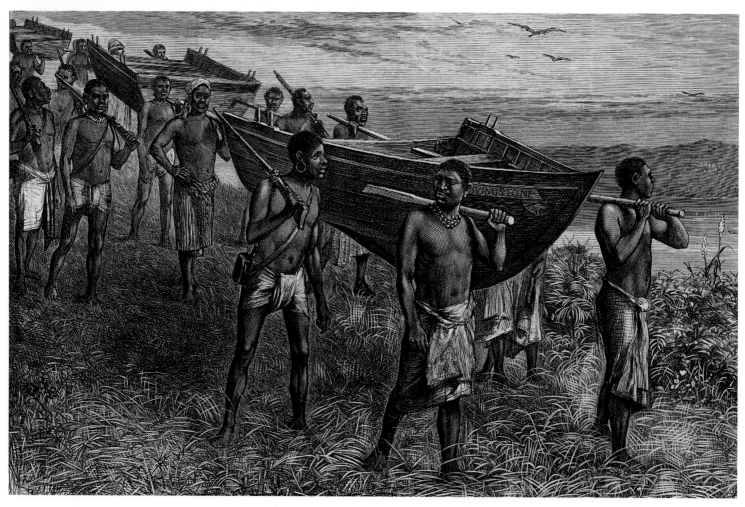

THE HEART OF DARKNESS

Well after the Niger and the Nile had been explored, the Congo Basin remained shrouded in mystery. Virtually the entire Congo watershed – more than a million square miles – was unknown. Tales of poisoned arrows and cannibals kept both Arab slavers and explorers away, and even David Livingstone shuddered at the thought of following the Lualaba River, west of Lake Tanganyika. He believed it was the headwater of the Nile, but was loath to confirm it as it might "turn out that I have been following the Congo, and who would risk being put into a cannibal pot ... for it?"

In 1816, a Royal Navy expedition under James Tuckey had been halted by the first of many cataracts on the Congo River, and within two months 35 of 54 expedition members, including Tuckey, had died of yellow fever. Thus, it was believed that following the Congo River into the interior would not be feasible.

The first European to penetrate deep into the Congo was Verney Lovett Cameron, a Royal Navy officer sent to bring supplies to Livingstone after he was "found" by Henry Morton Stanley. Cameron soon met Chuma and Susi bringing Livingstone's body back to the coast, so he decided to continue Livingstone's work. He made a thorough survey of Lake Tanganyika and then followed its only outlet, the Lukuga River, west to the Lualaba. Cameron hoped to travel north on the Lualaba, but he could not obtain boats and was forced to strike southwest into Katanga instead. In November 1875, he arrived at the Atlantic coast near Benguela in Angola,

having made the first crossing of central Africa.

Meanwhile, Stanley, who had become devoted to Livingstone, also decided to pursue the older man's researches. In 1874, sponsored by *The New York Herald* and *The Daily Telegraph* of London, he headed towards Lake Victoria with the most well-equipped expedition ever to leave the east coast. There, Stanley had his prefabricated boat *Lady Alice* put together, and confirmed both the extent of Speke's lake, and its discharge at Ripon Falls. He then marched to Lake Tanganyika, which he showed could not be related to the Nile. And then, refusing to be diverted like Cameron, he followed the Lualaba north into the Congo, navigated that river's great arc and portaged down its dozens of cataracts. In August 1877, 999 days after his departure, he reached the west coast.

Within several years, Stanley was back, working for Leopold II, the Belgian king, in the area south of the Congo River. Simultaneously, the French expanded into the region north of the river.

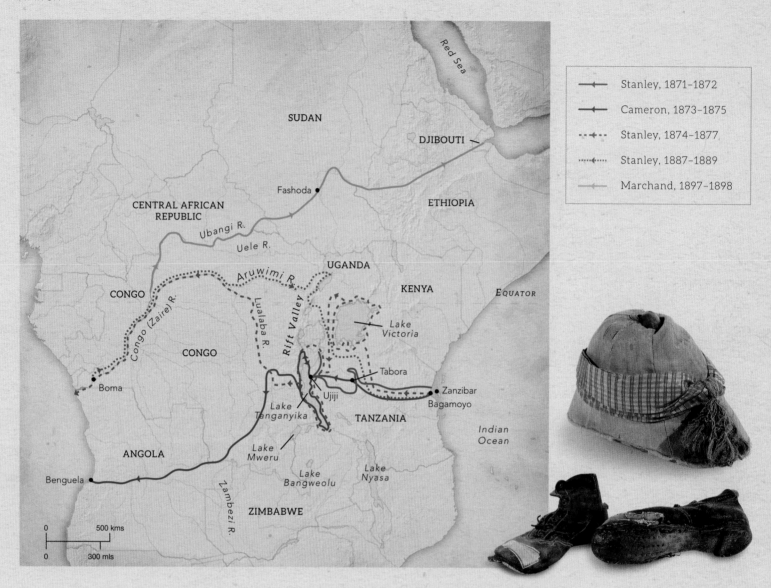

Legend:
- Stanley, 1871–1872
- Cameron, 1873–1875
- Stanley, 1874–1877
- Stanley, 1887–1889
- Marchand, 1897–1898

This effort was led by Pierre de Brazza, who in 1875–78 had explored the Ogowe and Alima rivers. Like Stanley, Brazza signed treaties with local African chiefs, and in November 1880 he founded Brazzaville, the capital of the French Congo.

Stanley's last great effort came on the Emin Pasha Relief Expedition. In 1881, an uprising under the Mahdi had slowly begun spreading across the Sudan, culminating in 1885 in General George Gordon's death in Khartoum. By that time, the Mahdi controlled virtually the entire Sudan except the southern province of Equatoria, where the governor Emin Pasha, a German convert to Islam, held out near Lake Albert. Stanley was chosen to rescue Emin, and in 1887, for various reasons his expedition departed from the Congo to cross Africa, rather than leaving from the much closer east coast. At the Aruwimi river, Stanley plunged into the unknown

FASHODA

By the mid-1890s the Mahdists still held sway in the Sudan, and the French decided to establish themselves on the upper Nile. In 1897–98, a contingent under Jean-Baptiste Marchand trekked from the Congo to the tiny Nile outpost of Fashoda, arriving seven weeks before a British army under Herbert Kitchener destroyed the Mahdist forces at Omdurman. Within days, Kitchener's gunboats headed upriver to force out the French. Marchand refused to leave, but the commanders sensibly passed any decisions to their governments. After a tense month, the French withdrew, leaving the Nile to the British.

TOP The hat that Henry Morton Stanley is reputed to have worn when he first met David Livingstone at Ujiji, near the shores of Lake Tanganyika.

ABOVE The boots Stanley wore all the way down the Congo on his first crossing of Africa (1874–77). All repairs were made by Stanley himself.

BELOW Two drawings by Henry Morton Stanley made during the Emin Pasha Relief Expedition. Many of Stanley's sketches were the basis for the artwork in his expedition account, *In Darkest Africa*.

Ituri rain forest. It had never been crossed by Europeans, and turned out to be sheer hell, perhaps the most dreadful ordeal any explorer of Africa ever faced. After six months, during which more than half of the party died and Stanley's hair turned white, they reached Lake Albert to find that Emin did not actually want to be rescued. Nevertheless, Stanley virtually forced Emin and his followers to accompany him back to the coast. On that final trek, Stanley made his last discovery: the Ruwenzori Mountains, in southwestern Uganda, completing the map linking the dark heart of Africa and the Nile.

LIEUT W·G·STAIRS R·E·
(BWANA M'SINGA = MASTER OF THE CANNON)

LEOPOLD'S CONGO FREE STATE

One of the most remarkable feats during "the scramble for Africa" was the establishment by Leopold II of Belgium of a personal empire. In 1879, Leopold sent Stanley to the Congo to sign treaties with local chiefs, build a railway up the cataracts and launch steamers on the upper river. By 1884, Stanley had accomplished all this, earning himself the nickname Bula Mutari, "Breaker of Rocks". For the next 25 years, Leopold made his Congo Free State a synonym for brutality and oppression, and mercilessly looted its resources until it was directly annexed to the Belgian state in 1908.

LEFT The first house built by Stanley in what would become Leopold's Congo Free State. Next to it are the machinery and materials to build a railway.

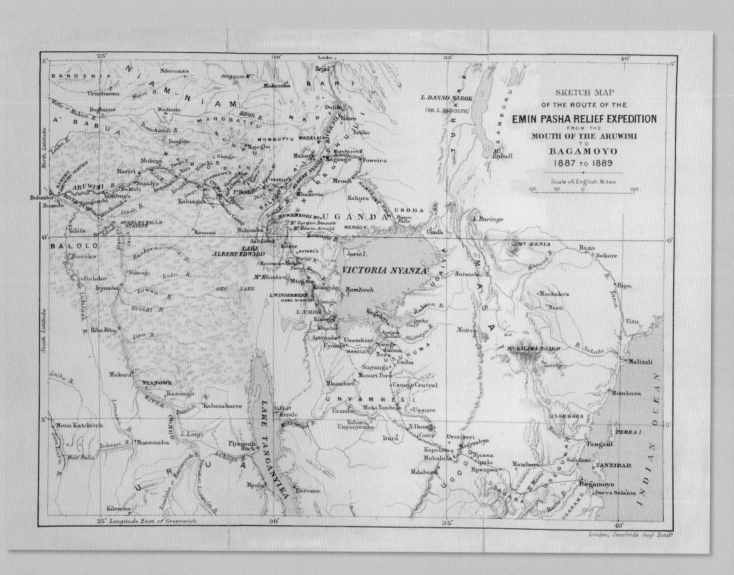

SKETCH MAP
OF THE ROUTE OF THE
EMIN PASHA RELIEF EXPEDITION
FROM THE
MOUTH OF THE ARUWIMI
TO
BAGAMOYO
1887 TO 1889

Scale of English Miles

ABOVE A map of part of the route taken by the Emin Pasha Relief Expedition. This was re-drawn for use at the RGS reception for Henry Morton Stanley after his return.

OPPOSITE Henry Morton Stanley, journalist and explorer, is possibly best remembered for his words "Dr Livingstone, I presume?" Also to his credit is the first circumnavigation of Lake Victoria Nyanza in 1875. At the time there were many differences of opinion as to the size of the lake. In a light boat called the Lady Alice, Stanley and 11 crew members travelled 1,000 miles around the lake in 57 days, thus conclusively proving its vast extent (it is the world's second-largest freshwater lake).

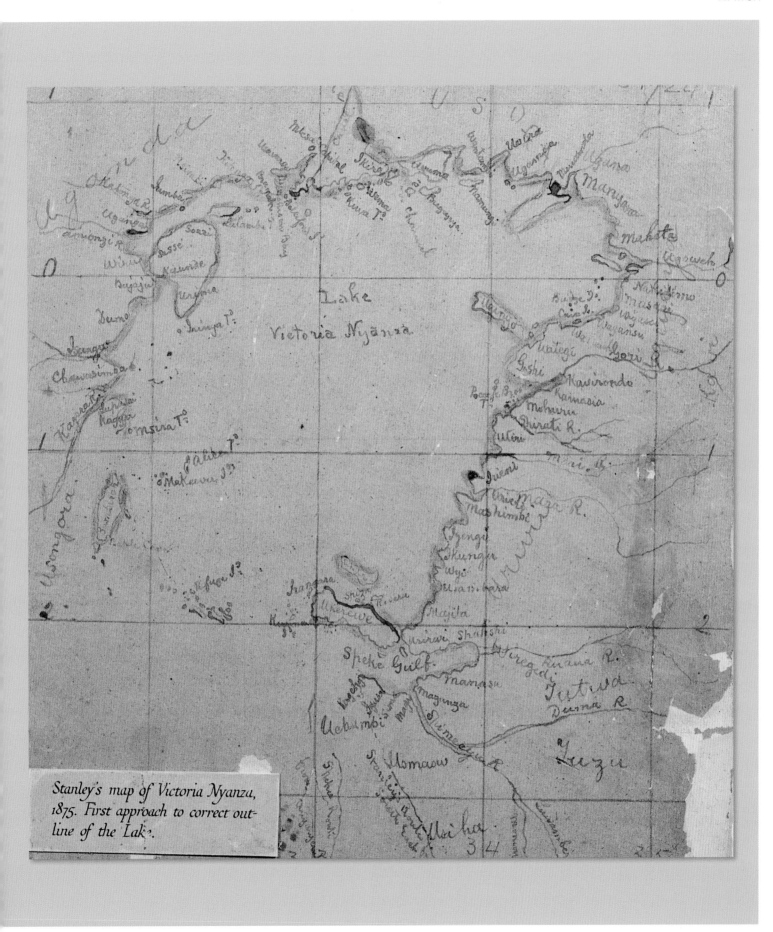

Stanley's map of Victoria Nyanza, 1875. First approach to correct outline of the Lake.

AUSTRALIA AND THE PACIFIC

A SECOND NEW WORLD

In 1605, soon after the Dutch East India Company had established a post at Bantam on Java, Willem Jansz was sent to find what other profitable trading areas might lie to the east. Reaching New Guinea, he followed its southern coastline before turning directly south. He crossed what he thought was an inlet and struck harsh, barren land. After sailing south along the coast, Jansz decided to turn north before his supplies ran short, and he made his way back to Java. Although he had become the first European to reach Australia, Jansz was unaware of this, for he did not know that the Torres Strait separated central New Guinea not from an adjoining peninsula, but from an entirely new continent.

OPPOSITE TOP LEFT William Dampier (1652–1715) was a true Renaissance man of the sea. He was a navigator, explorer, scientist, writer, and buccaneer.

OPPOSITE TOP RIGHT Matthew Flinders and George Bass in Bass' tiny boat *Tom Thumb*, in which the two made their first exploratory effort in 1795.

OPPOSITE BOTTOM Shark Bay on the coast of Western Australia, the point to which Tasman followed the coast and near which Dampier made landfall.

In the following decades, more voyages touched on the mysterious land the Dutch called New Holland, but it was not until 1642 that Anthony Van Diemen, governor of the Dutch East Indies, sent Abel Tasman to discover just how extensive it actually was. Unfortunately, Tasman sailed too far south before turning east, and therefore missed most of Australia, although he did reach an island he called Van Diemen's Land, a name later changed to Tasmania. Tasman then continued east and discovered New Zealand, before proceeding north to Tonga, Fiji and around the northern coasts of New Guinea. In 1644, Tasman made a second exploratory voyage, retracing Jansz's early route but following the northwest coast of Australia all the way to Shark Bay.

It was another five decades before Europeans made further significant contact with Australia. In 1688, while on a seven-year circumnavigation of the Earth, William Dampier landed in western Australia, one of the first Europeans to do so. Eleven years later, he was placed in command of the ship *Roebuck* by the Admiralty, and ordered to conduct scientific investigations in New Holland. Dampier sailed via the Cape of Good Hope and the Indian Ocean, reaching the continent's west coast in July 1699. He spent a month charting the coastline of what seemed to be a desolate and waterless area before sailing on. Worse was to come: *Roebuck* foundered on the return to Britain, and Dampier and his crew spent more than a month on Ascension Island awaiting rescue.

A century later, most of the Australian coastline remained unexplored, other than those parts charted

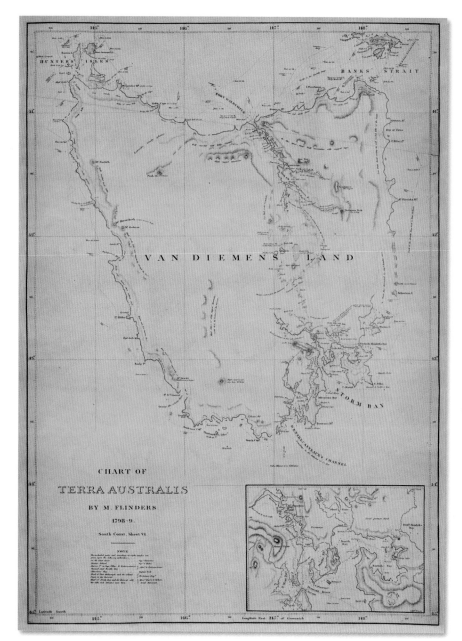

LEFT A map first produced by Matthew Flinders in 1798 during his coastal surveys of Van Diemen's Land, which proved it to be an island. He charted the north coast, entered the Tamar River, and proceeded around the western and southern coasts. He spent Christmas exploring the Derwent River.

ABOVE Two tools of the maritime trade in centuries past: an instrument known alternatively as dividers or a compass, and a telescope.

OPPOSITE Batavia (Jakarta today) was the administrative capital of the Dutch East Indies and the major port in the region for trade.

by James Cook in 1768–71. This was resolved to a great extent by one English and one French naval officer. Matthew Flinders first reached Australia in 1792 under the command of William Bligh. He returned in 1795 and, in the company of surgeon George Bass, conducted coastal exploration in New South Wales. In 1798–99, Flinders captained the tiny *Norfolk* and with Bass and a crew of eight made the first circumnavigation of Van Diemen's Land, proving it to be an island not a peninsula. In 1801, at the recommendation of Sir Joseph Banks of the Royal Society, Flinders was placed in command of the *Investigator* and sent to chart the entire coastline

ROBINSON CRUSOE BY ANY OTHER NAME

Alexander Selkirk was an adventurous young man from Largo, Scotland, when he joined a privateering voyage in 1703. The next year, having quarrelled with ship captain Thomas Stradling, he requested that he be put ashore on the nearby Juan Fernandez islands, off the coast of Chile. Selkirk remained on the island until 1709, when an expedition under Woodes Rogers found him, semi-wild, dressed in goatskins, and barely able to speak English. Daniel Defoe later used Selkirk's experiences as the basis for his classic novel *Robinson Crusoe*.

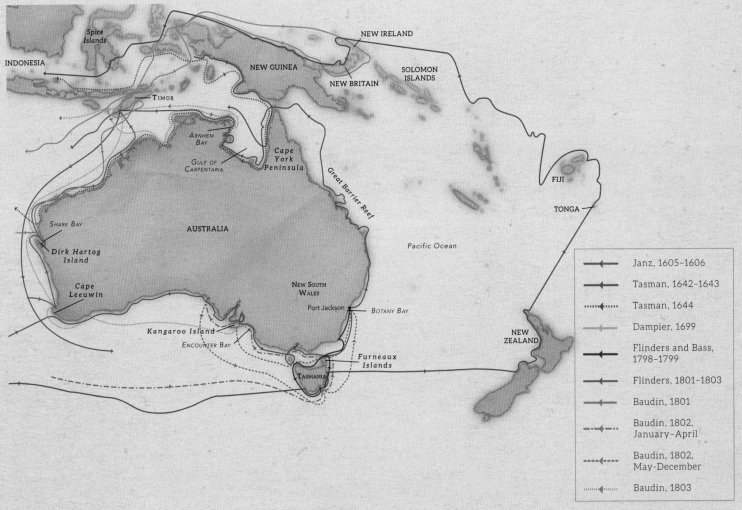

Legend:
- Janz, 1605–1606
- Tasman, 1642–1643
- Tasman, 1644
- Dampier, 1699
- Flinders and Bass, 1798–1799
- Flinders, 1801–1803
- Baudin, 1801
- Baudin, 1802, January–April
- Baudin, 1802, May–December
- Baudin, 1803

of Australia. Beginning at Cape Leeuwin on the southwest corner, he carefully mapped the south coast, along the way meeting the French expedition under Nicolas Baudin, with whom he generously shared his findings. Flinders thereafter mapped much of the east coast and the Gulf of Carpentaria before leaks and rotten planks forced him directly to Timor and thence onward to Port Jackson, which he reached in June 1803. He completed his circuit of Australia, although he was unable to finish his survey.

Meanwhile, Baudin had been placed in charge of two ships and directed to chart the Australian coast and, with 23 scientists, to compile extensive zoological and botanical collections. Arriving at Cape Leeuwin in May 1801, he spent two years making landfalls around the continent, although not circumnavigating it. Baudin died in Mauritius on his way home, and the French accounts of the expedition – written by two of the scientists – not only generally ignored his contributions, but did not acknowledge the charts Flinders had provided, claiming the discoveries for themselves.

THE DUTCH EAST INDIA COMPANY

In 1595, Cornelius Houtman made a pioneering Dutch commercial voyage to Java in the Spice Islands (now named Indonesia), infringing on a trade that had been dominated by Portugal and Spain. Soon numerous Dutch companies began trading in the region, and in 1602, they were amalgamated into the United Netherlands Chartered East India Company. The primary Dutch trading post became Batavia, and in the following decades the Dutch displaced the Portuguese and the Spanish. The company lasted until 1798, when the French conquest of Holland led to its demise.

EXPLORATION AND SCIENCE IN THE PACIFIC

At the close of the Seven Years' War in 1763, both the British and French navies had excess ships, officers and men. That, and the growing interest in science, led to a flurry of exploration in a little-known area: the Pacific. Within a year, John Byron – nicknamed "Foul-weather Jack" and the grandfather of the poet – was sent on a voyage of discovery. He claimed the Falkland Islands, not realizing that the Frenchman Louis-Antoine de Bougainville had already established a settlement there. Byron then made the quickest circumnavigation yet on record in 22 months.

OPPOSITE TOP Samuel Wallis is greeted by Queen Oberea of Tahiti in June 1867. Landing at Matavai Bay, Wallis stayed for a month, naming it King George Island and calling Moorea, its neighbour, York Island.

OPPOSITE LEFT An illustration of the landing of Comte de La Pérouse's party at Easter Island in April 1786. They remained for less than a day after giving the local inhabitants gifts of livestock, and orange and lemon trees.

OPPOSITE RIGHT As a midshipman in *Wager*, John Byron was shipwrecked on the Chilean coast in 1741. The ship's party divided in two, with Byron's group taking a small boat north until being picked up by a Spanish ship.

Upon Byron's return in 1766, another expedition was sent to extend his discoveries, under Samuel Wallis and Philip Carteret. After passing through the Strait of Magellan, Carteret's *Swallow* fell behind Wallis's vessel, and the two completed their voyages independently. Wallis crossed the Pacific and made landfall at Tahiti, an island never before visited by Europeans. He then sailed through the Society, Gilbert and Marshall islands, re-stocked in Batavia, and continued home without losing a single crewman. Carteret, meanwhile, discovered Pitcairn Island, but *Swallow* was damaged by violent storms, and the resulting lack of speed contributed to the death of 30 men from disease before the ship reached England 10 months after Wallis. The two had made limited discoveries, but had carefully charted large areas and laid the groundwork for James Cook.

Soon after Wallis and Carteret left England, Louis-Antoine de Bougainville sailed from Nantes. After handing the Falklands settlement to the Spanish, he, too, crossed the Pacific. In April 1768 he reached Tahiti, where the islanders saw through a disguise that had been maintained for an entire year: dressed as a man, the botanist's valet was actually a woman named Jeanne Barre. Then, in June 1768, Bougainville discovered the Great Barrier Reef, and, at New Ireland, oversaw measurements that established the correct longitude, allowing the exact measurement of the Pacific Ocean's width for the first time. The return to France completed the initial French circumnavigation, and made Barre the first woman to circle the globe.

After their participation in the American War of Independence, the French sent out further scientific expeditions. The most impressively equipped was that of Jean-François de Galaup, Comte de La Pérouse. Departing France in 1785, La Pérouse made one of the most geographically diverse examinations ever of the Pacific, following landings in Chile with voyages to Easter Island, Maui, Alaska, the North American coast and across the Pacific to China, Korea, Sakhalin and Kamchatka, before heading south to Botany Bay, where the British had just arrived with the first convicts. In early 1788, La Pérouse sailed north, never to be seen again.

In 1791, Joseph-Antoine Bruni d'Entrecasteaux was instructed to search for La Pérouse. He investigated many of the island groups near New Guinea and in the Dutch East Indies, and made a survey of sections of Australia's south coast. He then started a second broad sweep through the South Pacific islands, but the expedition's members were debilitated by dysentery and scurvy, and so d'Entrecasteaux decided to return to France, unaware that he had passed very near to Vanikoro, the site of La Pérouse's fatal shipwreck. Before long, d'Entrecasteaux himself died, and the ships were forced to call at Java, where the Dutch imprisoned many of the sailors for more than a year.

Meanwhile, the expedition that perhaps best typified the ideals and scientific hopes of the Enlightenment took place in 1789. Italian-born Alessandro Malaspina of the Spanish navy was sent to compile the most comprehensive hydrographic charts yet made, and to report on the political

and economic state of Spain's far-flung overseas possessions. During the next five years, he crossed and re-crossed the Pacific, examined the western coast of North and South America, visited China, the Philippines, the Dutch East Indies, Australia, New Zealand and islands throughout the Pacific. Following his return, Malaspina made strong political recommendations regarding the government of the Spanish colonies; some of these gained him powerful enemies who had him imprisoned. Tragically, his maps did not appear for 30 years, and a comprehensive edition of his journals was not published until 1990.

ABOVE The Great Barrier Reef off northeast Australia is one of the world's most remarkable maritime habitats. It was less appreciated by early sailors, who found its reefs incredibly dangerous.

KRUSENSTERN'S CIRCUMNAVIGATION

An Estonian-born officer of the Russian Navy, Adam Johann von Krusenstern proposed supplying Russia's American colonies via ships from Baltic ports rather than by the slow land-crossing of Siberia. Departing Kronstadt in mid-1803, Krusenstern entered the Pacific via Cape Horn, and reached Easter Island and Hawaii before proceeding to Alaska, the Aleutians and California. A visit to Japan failed to establish new trade agreements, but Krusenstern charted Japan's west coast and Sakhalin. He returned to Russia via Africa, having contributed greatly to knowledge of the Pacific and the North American coast.

LEFT Krusenstern, whose circumnavigation owed its inception in part to proposals of G. I. Mulovsky, who planned such an expedition but was killed before it took place.

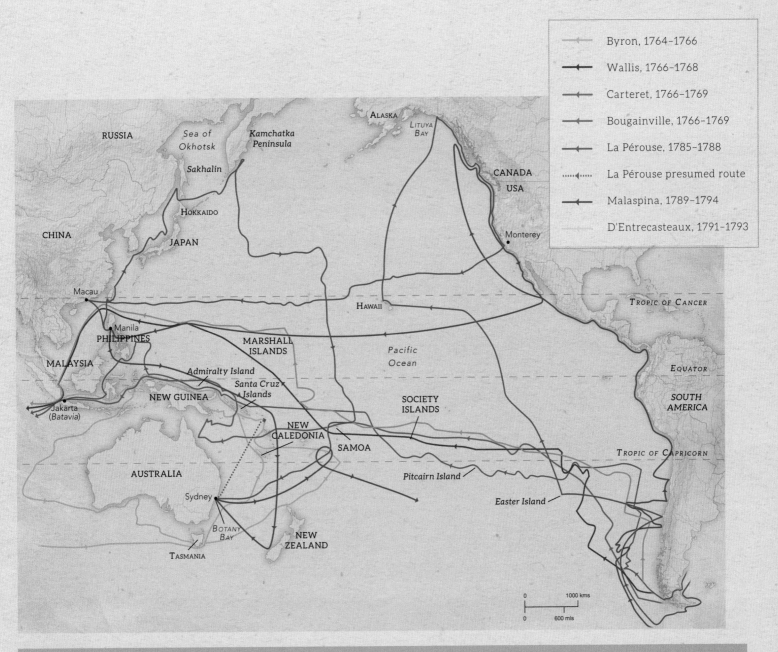

Legend:
- Byron, 1764–1766
- Wallis, 1766–1768
- Carteret, 1766–1769
- Bougainville, 1766–1769
- La Pérouse, 1785–1788
- La Pérouse presumed route
- Malaspina, 1789–1794
- D'Entrecasteaux, 1791–1793

THE FRENCH IN THE SUB-ANTARCTIC

French maritime exploration did not stop in the tropics. In 1738, Jean-Baptiste Bouvet de Lozier led an expedition to find the fabled great southern continent, "Terra Australis Incognita". Although unsuccessful, he did discover the most isolated island on Earth, today named Bouvetøya. Then, in 1771, Yves-Joseph de Kerguelen-Trémarec also went in search of the southern continent. He claimed to have found it, but when he returned to colonize it in 1773, it proved to be bleak, inhospitable islands later named Iles Kerguelen. Upon returning to France, Kerguelen was court-martialled and imprisoned for his false claims.

LEFT Kerguelen's claim that his discovery held "the very soil of France" led to his imprisonment in the fortress of Saumur.

135

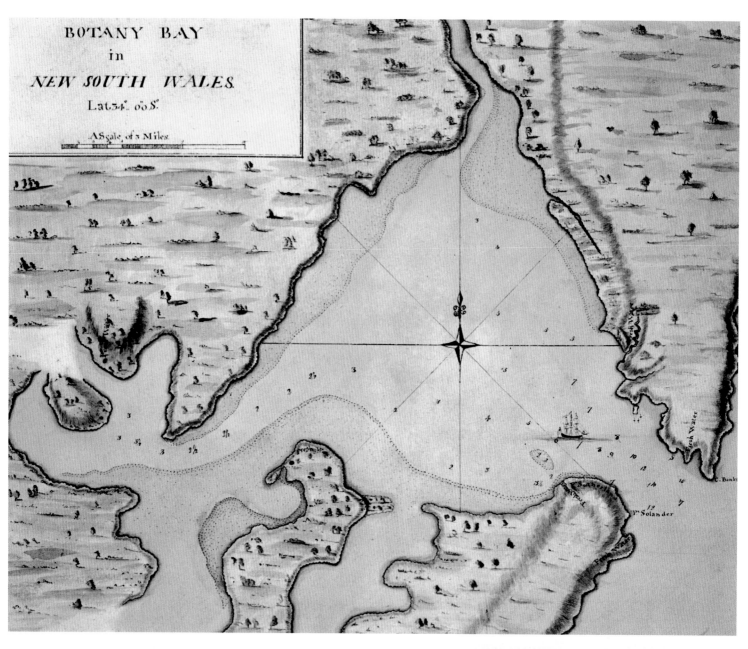

BOTANY BAY
in
NEW SOUTH WALES.
Lat.34°. 00 S.

A Scale of 3 Miles.

JAMES COOK

The distance of the Sun and other heavenly bodies from the Earth was one of the great scientific questions of the Enlightenment. It was believed that detailed observations of the transit of Venus – the occasions when that planet crossed the face of the Sun – would help determine these distances. Knowing that the transit of June 1769 would be the last one until 1874, the Royal Society proposed that observers should be sent to view it from Norway, Hudson Bay and the Pacific. Placed in command of the last of these expeditions was James Cook, a 39-year-old Royal Navy officer who had earned a reputation for his surveys of the St Lawrence River and the Newfoundland coast.

Samuel Wallis had just returned from his circumnavigation, and at his recommendation Tahiti was selected as the Pacific site. Cook sailed on HMS *Endeavour* in August 1768, carrying a large complement of scientists – including the botanist Joseph Banks – and holding secret orders instructing him to seek the legendary southern continent, "Terra Australis Incognita". The transit was successfully observed from Tahiti, following which Cook sailed south. However, finding no land, he headed to New Zealand and spent six months charting its coasts. Cook then made the first recorded sighting of eastern Australia, landed at Botany Bay on 28 April 1770, and sailed north through the Great Barrier Reef. After confirming the existence of a strait between Australia and New Guinea, he returned to England in July 1771.

The next year, Cook was sent on a more extensive search for the southern continent. Commanding *Resolution* and *Adventure*, he ploughed south from Cape Town, making the first crossing of the Antarctic Circle on 17 January 1773. Having been forced back by ice, he unsuccessfully searched for Iles Kerguelen before stopping in New Zealand and Tahiti. He then turned south again, attaining a record latitude of 71° 10' S before reaching more impenetrable pack ice. Yet another sweep through the South Pacific allowed him to rediscover the Marquesas, visit Easter Island, and discover Norfolk Island. He then entered high southern latitudes yet again, sailing south of Cape Horn, discovering the South Sandwich Islands and claiming South Georgia.

In July 1776, just a year after reaching Britain, Cook departed on a third voyage, to investigate the northwest coast of America and determine

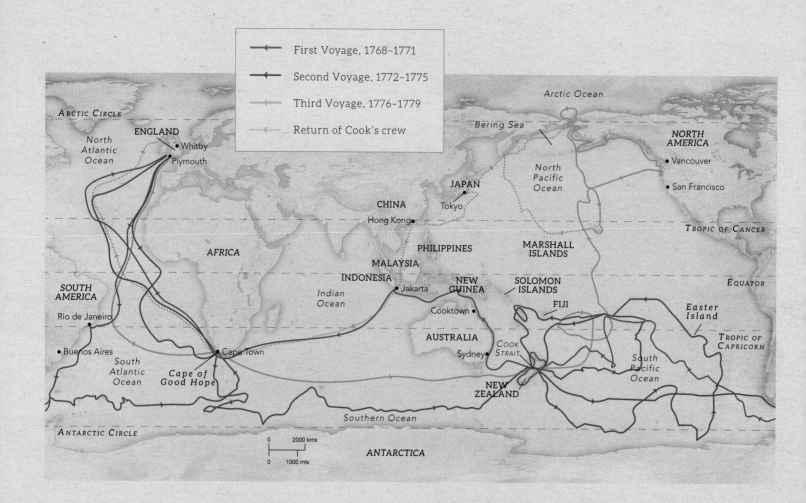

First Voyage, 1768–1771
Second Voyage, 1772–1775
Third Voyage, 1776–1779
Return of Cook's crew

conclusively whether there was a Northwest Passage. Sailing southeast from Cape Town, this time he did locate Iles Kerguelen, which he charted before proceeding to Van Diemen's Land and New Zealand. He then discovered Raratonga and the Cook Islands before sailing north and discovering the Hawaiian Islands, which he named the Sandwich Islands. The ships progressed to North America, following the coast from Oregon to Alaska, although no inlets that might be the mysterious passage were found. Passing through the Bering Strait, the party was stopped by heavy ice at 70° 44' N, but landed on the Asian coast before heading for Hawaii to winter.

Although relations with the Hawaiians started positively, they ultimately became strained, and Cook decided to leave in February 1779. However, damage to the mast of *Resolution* forced a return, and while repairs were being carried out, one the boats of *Discovery* was stolen. Cook went ashore

SCURVY

For centuries, scurvy was the greatest menace to long-distance voyagers. Caused by a deficiency of vitamin C, scurvy first appears as a swelling of the gums and loosening of the teeth, and untreated it can lead to death. Sailors or surgeons often found cures – such as eating fresh fruit or vegetables – but their discoveries were then lost due to numerous confounding factors. Captain Cook kept his crews relatively scurvy-free by restocking regularly and insisting they eat a wide variety of items, some of which contained vitamin C.

ABOVE An artist's impression of the killing of Cook at Kealakekua Bay, Hawaii. The usually even-tempered Cook had been unusually volatile in the later stages of his last voyage, and it showed when he tried to recover the stolen boat.

with a party of marines to recover it, but in the resulting fracas he was killed. The expedition continued under the second-in-command, Charles Clerke, who once again sailed through the Bering Strait. The expedition progressed no farther than on the previous attempt, however, and upon retreating to Petropavlovsk, Clerke died of tuberculosis in August 1779 aged 36. The ships returned to England under new captains.

In his three magnificent expeditions Cook had discovered many new lands, mapped vast swathes of previously unknown waters, overseen the advancement of scientific knowledge, laid to rest the notion of a great southern continent in a temperate region and showed it was possible to hold scurvy at bay even during multi-year voyages. His contributions are unsurpassed in the history of exploration.

JOSEPH BANKS

Although he is forever linked with Cook, Joseph Banks was himself actually the inspiration behind many scientific expeditions. Elected to the Royal Society at only 23, two years later he accompanied Cook's first voyage to study botany, after personally contributing much scientific equipment. He planned to sail on Cook's second expedition but withdrew after difficulties about the size of his entourage, instead visiting Iceland. In 1778, he became president of the Royal Society, serving for 42 years and advising expeditions led by William Bligh, Constantine Phipps, George Vancouver, Matthew Flinders and William Scoresby.

LEFT Joseph Banks was the primary force in founding the African Association, the virtual director of Kew Gardens for decades, and a key figure in the establishment of the colony of New South Wales.

INTO THE AUSTRALIAN INTERIOR

Settling in the fledgling colony of New South Wales proved harder than anyone had imagined, and so it was several decades before any attention was given to the exploration of the Australian interior. In 1813, Gregory Blaxland and two colleagues became the first to cross the Blue Mountains, and their reports led Governor Lachlan Macquarie to establish a road and the first significant inland settlement, Bathurst. In the ensuing years, John Oxley, surveyor-general of New South Wales, and botanist Allan Cunningham made journeys north and northwest, discovering the Macquarie Marshes and the Liverpool Plains.

OPPOSITE TOP An illustration of the citizens of Adelaide turning out in August 1844 to see Sturt off on his attempt to reach the continent's centre. Sturt actually mustered his full party at Moorundi on the Murray River.

OPPOSITE BOTTOM The beautiful Blue Mountains of New South Wales. Their jagged ravines, high ridges and dense undergrowth made them seem impassable to the early settlers. But in 1813 Gregory Blaxland, William Wentworth and William Lawson proved they could be conquered.

RIGHT Thomas Mitchell's series of journeys into the interiors of what are now New South Wales and Victoria, along with his continuing work as surveyor-general of New South Wales, led to him being knighted in 1839.

BOTTOM RIGHT Perhaps the greatest explorer of Australia, Charles Sturt. At differing times Sturt served as South Australia's surveyor-general, assistant commissioner of land, registrar-general and colonial treasurer. But what he most loved was exploring unknown lands.

The first major explorer of the interior, however, was Charles Sturt, an army officer who was sent in 1828 to explore the Macquarie River. His second-in-command was Hamilton Hume, who had earlier led a party towards the south coast, reaching it just west of present-day Melbourne. Sturt and Hume followed the route pioneered by Oxley but carried on into the Macquarie Marshes and thence along the Darling River before retreating because the water was too salty to drink.

The next year Sturt was off again, this time west along the fast-flowing Murrumbidgee River in an 8-metre (27-foot) whaleboat. After a week, the party entered the broad Murray River, then, after narrowly avoiding attack by hundreds of Aborigines, passed a junction with what Sturt thought to be the Darling. They continued their 1,600-kilometre (1,000-mile) outbound journey, reaching the ocean, near present-day Adelaide. The return was a nightmare, but they reached Sydney safely, and the expedition helped open up South Australia, as did the political influence of Edward Gibbon Wakefield, a theorist on colonization who had already suggested new methods of land distribution in New South Wales.

Sturt's explorations were extended by Oxley's successor as surveyor-general, Thomas Mitchell, who in 1831 had already led an expedition northwards that was aborted when several of its members were killed by Aborigines. In 1835, Mitchell traced the Darling 475 kilometres (300 miles) southwest from the marshes, turning back before reaching the Murray due to a skirmish with Aborigines. The next year, he followed the Lachlan River to the Murrumbidgee and thence up the Murray, confirming that Sturt had been correct in supposing that the Darling flowed into it. Mitchell then turned east and south, continuing to the coast near Portland Bay. Trying to make a straight line to Sydney, he passed through some exceptionally fertile areas. Mitchell's reports encouraged many settlers to follow his tracks into what became Victoria.

Meanwhile, more routes were being blazed into South Australia. Edward Eyre played a key role in this process, making a series of cattle and sheep drives from New South Wales. Fired by encounters with Sturt, Eyre began exploring northwest of Adelaide, and in 1839 he discovered the huge salt lake he named Lake Torrens. The next year, he discovered another salt lake farther north, which he thought was part of the first, but which would later be

named Lake Eyre. Eyre then tried to reach Western Australia, following the coast with his friend John Baxter. But in April 1841, Baxter was murdered by two Aborigines, who stole the remaining food. Eyre and an Aborigine named Wylie continued and, with the assistance of a French whaling ship, made it to Albany, having walked almost 1,600 kilometres (1,000 miles). They were the first men ever to reach Western Australia overland.

BELOW Edward Eyre greets Captain Rossiter of the French whaler *Mississippi* as Wylie waves to the crewmen. The explorers, nearly on their last legs, rested aboard ship for thirteen days before continuing their journey along the southern coast.

TRANSPORTATION

The loss of the American colonies after the American War of Independence, left the British policy of "transportation" – sending criminals overseas rather than to prison – in disarray. The mistakenly favourable reports about Botany Bay by Captain Cook and Joseph Banks suggested an alternative solution. What is known as the "First Fleet" arrived in Botany Bay in January 1788 with approximately 1,000 people, three-quarters of them convicts. Governor Arthur Phillips soon shifted the colony to Port Jackson, the site of present-day Sydney. Despite conditions more gruelling than expected, 160,000 convicts were transported to Australia before the last shipment in 1868.

RIGHT The ships of the First Fleet in January 1788, when they unloaded some 730 convicts, plus sailors, marines, livestock and supplies for the new colony.

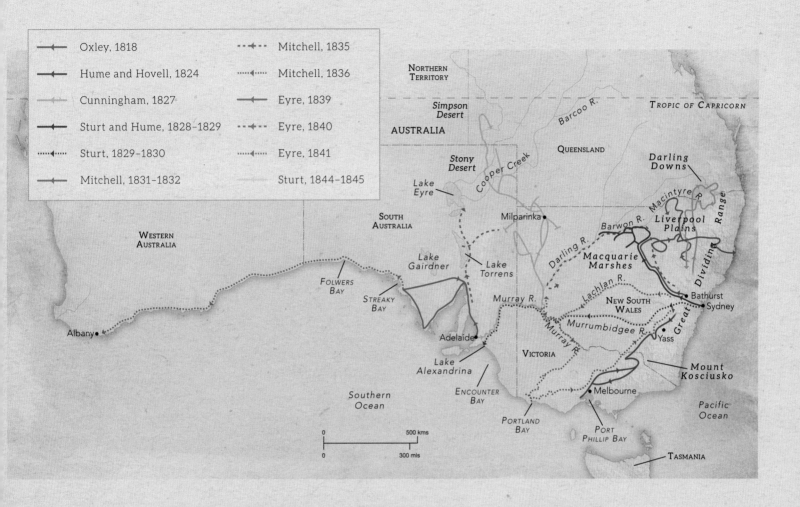

Legend:
- Oxley, 1818
- Hume and Hovell, 1824
- Cunningham, 1827
- Sturt and Hume, 1828–1829
- Sturt, 1829–1830
- Mitchell, 1831–1832
- Mitchell, 1835
- Mitchell, 1836
- Eyre, 1839
- Eyre, 1840
- Eyre, 1841
- Sturt, 1844–1845

THE NEW ZEALAND COMPANY

In 1837, Edward Gibbon Wakefield was the key thinker behind the establishment of what became the New Zealand Company, which Wakefield hoped would oversee the orderly (but profitable) settlement of New Zealand. Although Wakefield's plans were not approved by the British government, in the 1840s the Company sent explorers to chart the islands and investigate settlement possibilities. Among the most significant of these were Charles Heaphy, who explored both North and South Islands, and Thomas Brunner, who, along with Heaphy, discovered Lake Rotoroa and made a pioneering journey down the South Island's west coast.

Three years after Eyre's return, another expedition left Adelaide. Sturt hoped to verify the existence of a mountain range in the heart of the continent, find the great inland sea he believed existed and establish new grazing grounds. After ascending sections of the Murray and the Darling, Sturt pressed towards the centre of the continent, but was soon forced to remain at a waterhole as the surrounding rivers dried up. After six months there, rains allowed the party to move on, and they pushed 650 kilometres (400 miles) northwest through the Stony Desert and past Cooper Creek to the edge of the Simpson Desert. Confronted by this impassable obstacle, and with scurvy debilitating the party, Sturt retreated, struggling back into Adelaide in January 1846.

While this exploration was going on, Wakefield remained first and foremost concerned with colonization, but the difficulties he encountered in Australia led him to find a new place to test his theories: New Zealand.

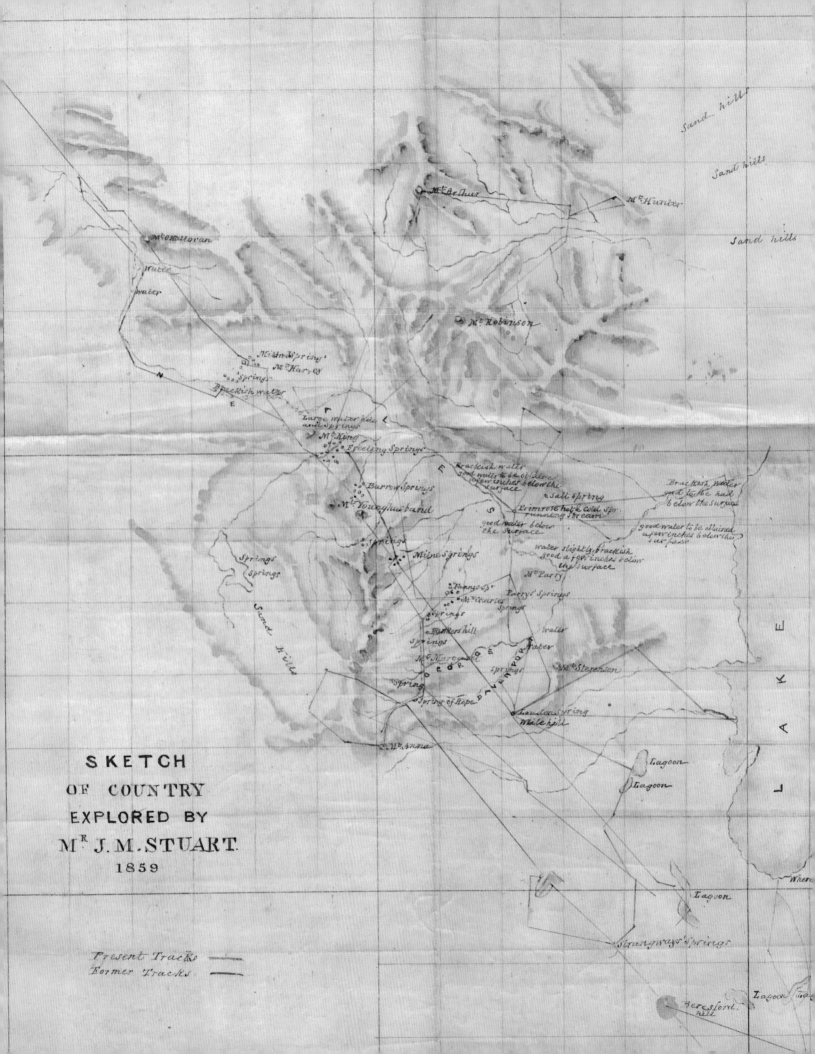

Sand hills

Sand hills

Sand hills

Mc Arthur

Mc Hunter

M^c O'Halloran

Water

Water

Mt Robinson

Mc Harvey Mildred spring

Spring

Brackish water

Large water hole and springs

Mc King

Freeling Springs

Brackish water good water to be obtained a few inches below the surface

Salt spring

Brackish water good to the head 6 below the surface

Barrow Springs

Primrose hole cold spr. running stream

good water below the surface

good water to be obtained a few inches below the surface

Mt Younghusband

Springs

Milne Springs

water slightly brackish good a few inches below the surface

Mt Parry

Springs

Springs

Fanny Spr.

Mc Charles

Parry's Springs

Sand Hills

Hunters hill

Water

Springs

Water

Mt Stevenson

Mc Margaret

Springs

Spring

Spring of Hope

Loudon spring

Milk hill

Mc Anna

Lagoon

Lagoon

LAKE

S K E T C H
OF COUNTRY
EXPLORED BY
MR J. M. STUART.
1859

Lagoon

Where

Strangways Springs

Beresford hill

Lagoon

Strangways Springs

Present Tracks ————
Former Tracks ————

ACROSS THE CONTINENT

By the mid-1850s, much of the eastern third of Australia had been explored, but the parched centre remained unknown. In 1855, the British government, via the Royal Geographical Society, sponsored Augustus Gregory to investigate this harsh region while also searching for the missing Ludwig Leichhardt. Gregory, whose reconnaissance north of Perth in the 1840s had helped expand western agricultural areas, landed by ship on the north coast and advanced south several times, hoping to cross the continent. Each time his party was forced back by lack of water, and eventually they turned east to Queensland, following a course similar to that of Leichhardt's major expedition. Two years later, Gregory made another unsuccessful search for Leichhardt, concurrently proving that the Barcoo River was linked to Cooper Creek.

OPPOSITE Sketch of the country around Lake Torrens, Australia, in 1859. Australian explorer John McDouall Stuart was born in Scotland and moved to Australia when he was 23. Between 1855 and 1862 he made six expeditions north from Adelaide into the heart of Australia. On a trip in 1859 he explored the area to the north of Lake Torrens, the lake shown in his sketch. During that trip, seven artesian springs were discovered by the German botanist, Joseph Albert Herrgott, who was travelling with him. Stuart named them Herrgott Springs, a name subsequently changed to Marree.

ABOVE Although John McDouall Stuart did not cross the continent as quickly as Burke and Wills, he lived to tell the tale. Unfortunately, his health was severely damaged, and he never truly recovered.

RIGHT TOP John O'Hara Burke, a native Irishman, emigrated to Australia in 1853 and served as a local police officer. However, he had little background to become an efficient explorer.

RIGHT BOTTOM William Wills was a trained surveyor from Devon, who moved to Australia in 1852. His diary and letters supplied much of the information about what happened on the expedition.

Such lack of success only spurred more interest, and in 1860, two south–north crossings were attempted. In March, John McDouall Stuart, who had participated in Sturt's last expedition, set out from Adelaide. He passed through the Macdonnell Range, reaching a landmark he named Central Mount Sturt, but that was later renamed after Stuart. He then pushed on for another 325 kilometres (200 miles) before being confronted by Aborigines at what he named Attack Creek. As he was running out of supplies, Stuart turned back, reaching Adelaide in early October.

Meanwhile, in August, to great fanfare, another trans-continental expedition left Melbourne. Sponsored by the Victoria government, scientific societies and wealthy individuals, it was led by a local policeman, Robert O'Hara Burke, who took command after Gregory declined the position. Burke's party of 18 was equipped with horses, camels, and 21 tons of equipment, but suffered in its leadership – Burke made several crucial blunders. First, he took an advance party to Cooper Creek without giving clear instructions to a second party about bringing

up the main supplies, which never arrived. Then, at Cooper Creek, worried Stuart might win the race, Burke again divided his party. He took three men – including his deputy William Wills – six camels, and three months' of supplies for a dash north. After two months, the group reached the hinterlands of the Gulf of Carpentaria, but massive mangrove swamps prevented them from attaining the coast proper. They then headed back, with awful travelling conditions compounded by a lack of food and water. One man died, and when, four months later, the surviving three stumbled into their base, they found the support team had given up and left that very morning. They tried to reach a settlement

LUDWIG LEICHHARDT

Born in Prussia in 1813, Ludwig Leichhardt emigrated to Australia in 1842 and soon began conducting short scientific expeditions in New South Wales. In 1844, he headed northwest from Brisbane to establish a land link with Port Essington on the Cobourg Peninsula. Despite undoubted scientific abilities, Leichhardt lacked both leadership and navigation skills, once calculating his position as 32 kilometre (20 miles) out at sea. Nevertheless, in December 1845 he reached Port Essington. Leichhardt then planned the first east–west crossing of Australia, but after setting out in April 1848, he and his party disappeared.

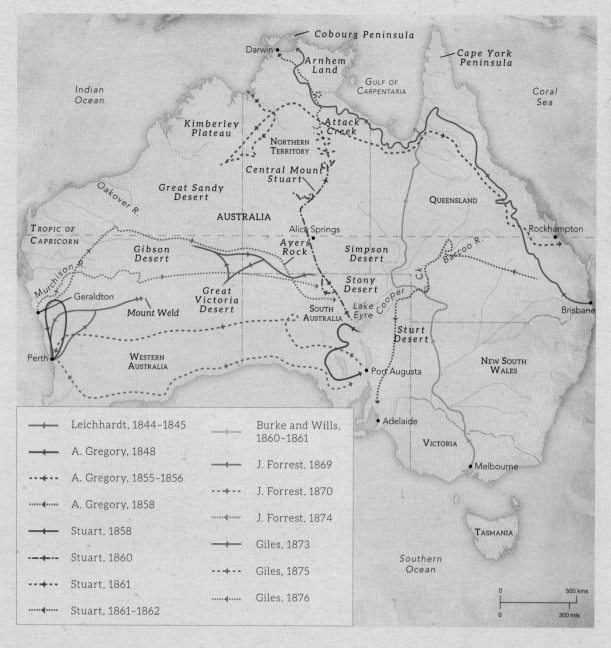

ABOVE Carving on the tree at Burke's stockade, which told the explorers where a small food cache had been buried when their support party had left that very morning. It was not enough to save their lives.

Leichhardt, 1844–1845	Burke and Wills, 1860–1861
A. Gregory, 1848	J. Forrest, 1869
A. Gregory, 1855–1856	J. Forrest, 1870
A. Gregory, 1858	J. Forrest, 1874
Stuart, 1858	Giles, 1873
Stuart, 1860	Giles, 1875
Stuart, 1861	Giles, 1876
Stuart, 1861–1862	

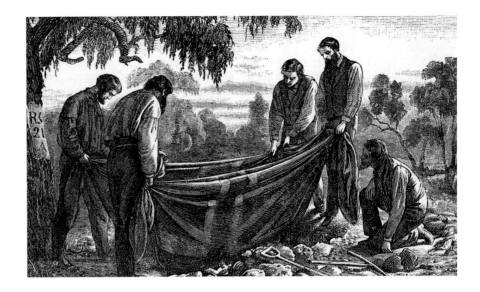

ABOVE In later life, John Forrest served as the Premier of Western Australia for 10 years. In 1918 he became the first native-born Australian to be raised to the peerage, as Baron Forrest of Bunbury.

ABOVE RIGHT A search party under Alfred Howitt found the bodies of Burke and Wills and rescued John King, who was on the verge of death despite generous assistance from Aborigines.

240 kilometres (150 miles) away, but Burke and Wills both died of starvation. The final man, John King, was found two months later by a rescue party, having been kept alive by Aborigines.

Stuart, meanwhile, returned in early 1861 to complete his own crossing. But after following his earlier path to Attack Creek, he could advance only about 160 kilometres (100 miles) through dense scrub before again retreating. Then, in October, he finally forced his way through the badlands and reached the coast near Darwin. Stuart had not only crossed the continent, but had pioneered the route of the Central Overland Telegraph Line.

With the trans-continental crossing completed, the exploration of Western Australia now came to the fore. John Forrest led three major expeditions, the first (in 1869) a 3,200-kilometre (2,000-mile) trek from Perth in search of agricultural land. In 1870, he made the first west–east crossing of the region, reversing Edward Eyre's trail, and in 1874 he again crossed Western Australia, this time from Geraldton to the line of the Central Overland Telegraph. Almost simultaneously, Ernest Giles was also crisscrossing the western deserts.

In 1873, Giles attempted an east–west journey; his colleague died, and Giles only just lived to tell the tale, but this did not put him off. In 1875, he crossed the Great Victoria Desert on a 4,000-kilometre (2,500-mile) march from Port Augustus, South Australia, to Perth. After resting several months, he returned, following a new route north of Forrest's second one. Having barely survived, Giles proclaimed his triumph complete: there was, he said, nothing left in Australia to be discovered.

CAMELS IN THE OUTBACK

The scarcity of water in the interior of Australia made it every bit as difficult and dangerous for horses as for men. However, camels can go for weeks without drinking and can survive on vegetation that would not support other pack animals, so it was not long before Australian explorers began to import them, an early example being John Horrocks in 1846. Burke imported camels from India for his expedition, and they have since been used frequently, including in 1993 when famed polar adventurer Geoff Somers crossed Western Australia with a pair.

RIGHT For his later expeditions, Ernest Giles used camels as his pack animals. Without them he certainly would have perished on these gruelling treks.

THE ARCTIC

LANDS OF EVERLASTING SNOW AND ICE

Defining Arctic exploration has been made significantly more difficult by the fact that what exactly is considered to constitute "the Arctic" has varied enormously through the ages – and continues to differ even today. For example, neither Pytheas – the fourth-century BC Greek navigator often mentioned as the earliest Arctic explorer – nor the numerous sixteenth- or seventeenth-century Northwest Passage expeditions ever penetrated beyond what today is regarded as the sub-Arctic, despite their reports of uninhabitable lands and ice-covered seas. Similarly, through time, what was deemed "the high Arctic" actually moved northward as previously remote, hostile areas were explored and became familiar.

OPPOSITE This early map of the Arctic or Northern Lands was drawn by Flemish cartographer Gerard Mercator. While aspects of this map represent a modern view of the world – its periphery was drawn from existing maps based on the latest information from English and Dutch explorers – the heart is mediaeval. At the North Pole stands a mythical magnetic mountain surrounded by a large landmass divided by four rivers. These rivers were said to draw ships into a whirlpool around the mountain. The circular maps in three corners show the Shetland and Faroe Islands and the mythical island of Frisland. The term "septentrional" as shown in the fourth corner is used because the Northern Lands were associated with the seven stars in the constellation Ursa Major.

RIGHT A dangerous moment for a ship piloted by Willem Barentsz through the exceptionally icy seas of the Arctic in the 1590s.

The voyage of Pytheas did, however, explore areas previously unknown to the Greek world. Sailing from Massilia (Marseille) around 325 BC, he probably circumnavigated Britain, and also claimed to have seen the land of "Thule" near the Arctic Circle. There has been considerable debate about what precisely Thule was: whether the Shetlands, Faeroes, Iceland or Norway. Another early voyage was that made by the Norseman Ottar, who reported to Alfred the Great that around AD 880 he sailed northeast from northern Norway, discovering extensive walrus-hunting grounds. This probably constituted the first recorded voyage into the Barents and White seas.

Seven centuries later, that region ignited the interest of a group of London merchants who sponsored an attempt to navigate a north-east passage to Cathay. In 1553, Sir Hugh Willoughby led three ships to the Barents Sea before losing contact with one under Richard Chancellor. Willoughby sailed on east to Novaya Zemlya, but was then forced by ice to retreat. Trapped on the Kola Peninsula for the winter, the entire party of some 66 men died of cold and starvation. Chancellor, meanwhile, sailed into the White Sea, was guided on to Moscow, and made trade arrangements with Tsar Ivan IV, leading

to the formation of the Muscovy Company in February 1555. The same year, Chancellor returned to Moscow, but then lost his life when he was shipwrecked off Scotland on the return voyage.

In 1556, Stephen Borough was sent by the Muscovy Company to explore to the east of the White Sea. He proceeded to Ostrov Vaygach, south of Novaya Zemlya, before being stopped by ice. The following year, Anthony Jenkinson made his great journey from Moscow far south to Bukhara. This led to the Muscovy Company's initiation of expeditions on this more southerly route to the Orient, but the company also continued to send ships via the White Sea. In 1580, Arthur Pet and Charles Jackman became the first Europeans to pass Novaya Zemlya and sail into the Kara Sea. Jackman's ship, however, disappeared on its return.

ARTHUR DOBBS AND THE NORTHWEST PASSAGE

Arthur Dobbs, an influential expert on Irish and colonial economics, was also obsessed with the Northwest Passage, which he insisted opened into Hudson Bay. Pushed into action by his politicking, the Admiralty sent Christopher Middleton to investigate in 1741. Middleton found no passage, but his conclusions were publicly attacked by Dobbs, who claimed Middleton and the Hudson's Bay Company were concealing the truth. In 1746 Dobbs organized a private expedition under William Moor, whose careful searches effectively ended Dobbs's long campaign. Dobbs later became governor of the colony of North Carolina.

ABOVE An example of the endless coast in the Canadian High Arctic, seen through open waters that become iced over during the winter.

HORATIO NELSON AND THE BEAR

In 1773, the Admiralty sent two ships under Constantine Phipps to reach the North Pole. The expedition attained a record farthest north of 80° 48' N, but is better remembered for an incident involving 14-year-old Horatio Nelson, then a midshipman. One night, while the ship was caught in the ice, Nelson was suddenly seen far away attacking a huge bear. Nelson's musket flashed in the pan, so he went after the bear with the butt. The captain fired a cannon, which frightened off the bear, and Nelson returned grumbling that he wanted the bearskin for his father.

LEFT The most famous visual representation of Nelson and the bear was Richard Westall's 1809 painting, which was then engraved by John Landseer to appear in James Clarke and John McArthur's classic two-volume biography *The Life of Admiral Lord Nelson*.

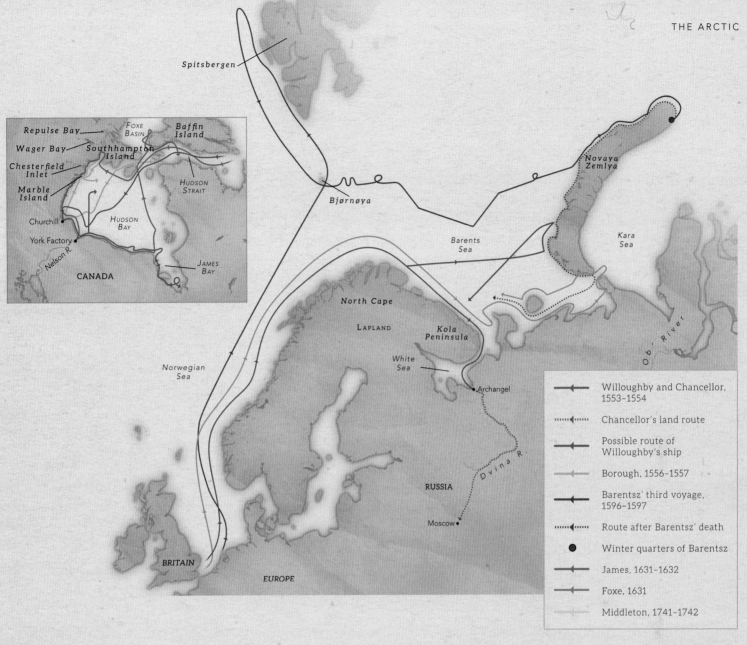

Spitsbergen

Repulse Bay | Foxe Basin | Baffin Island
Wager Bay | Southhampton Island
Chesterfield Inlet
Marble Island | Hudson Strait
Churchill
York Factory | Hudson Bay
Nelson R. | James Bay
CANADA

Novaya Zemlya

Bjørnøya

Kara Sea

Barents Sea

North Cape

LAPLAND

Kola Peninsula

White Sea

Norwegian Sea

Ob' River

Archangel

Dvina R.

RUSSIA

Moscow

BRITAIN

EUROPE

→	Willoughby and Chancellor, 1553–1554
◄····	Chancellor's land route
←	Possible route of Willoughby's ship
←	Borough, 1556–1557
←	Barentsz' third voyage, 1596–1597
◄····	Route after Barentsz' death
●	Winter quarters of Barentsz
←	James, 1631–1632
←	Foxe, 1631
←	Middleton, 1741–1742

OPPOSITE An account written by Gerrit de Veer, an officer serving with Barentsz, is the key source for information about those voyages. This engraving from his chronicle shows an attacking polar bear being fired upon.

Meanwhile, the Dutch followed the British lead into the White Sea and the Northeast Passage. In 1594, and again the next year, ships piloted by Willem Barentsz reached Novaya Zemlya but could not penetrate the Kara Sea due to heavy ice. So in 1596, two ships with Barentsz as pilot took a more northerly route, thereby discovering Bjørnøya and making the first confirmed sighting of Spitsbergen. The two ships then parted, with Barentsz's continuing east and sailing north around Novaya Zemlya before being trapped by the ice and forced to spend a miserable winter there. Unable to release the ship, the men took open boats to the Russian mainland, where they were rescued by Russian fishermen, but only after five of them, including Barentsz, had died of scurvy and the intense cold.

The Dutch also employed Henry Hudson to seek a Northeast Passage, although he eventually reversed directions and searched instead in northern Canada, discovering Hudson Bay on his final voyage in 1610. Hudson's successors in North America included Luke Foxe and Thomas James, who were sent in 1631 on rival expeditions by London and Bristol merchants, respectively, to seek out a Northwest Passage. They independently extended the discoveries of Hudson and Thomas Button (see page 41), while also showing that Hudson Bay had no western outlet. This effectively ended the British search for the Northwest Passage until 1719, when the Hudson's Bay Company sent James Knight to search the west coast of Hudson Bay for the legendary Strait of Anian. But Knight's expedition vanished, and it was not until the 1760s that it was discovered they had all died within two years of sickness and famine at Marble Island.

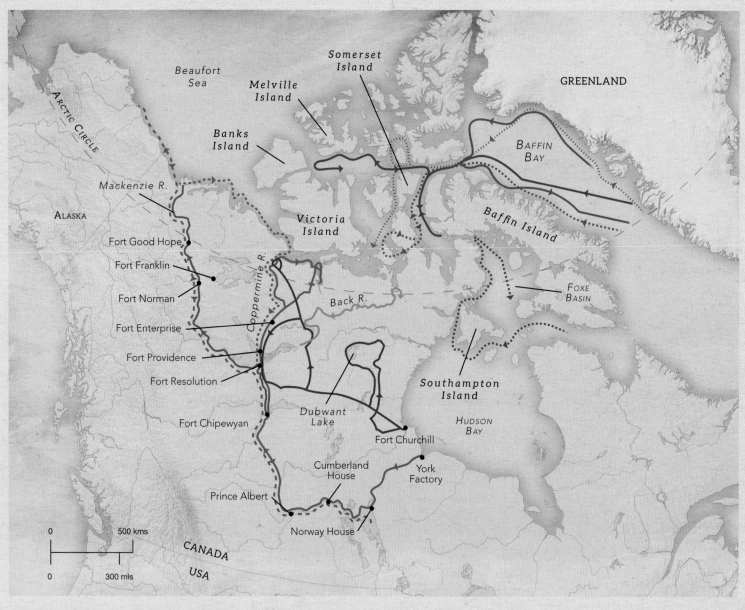

Beaufort
Sea

ARCTIC CIRCLE

*Somerset
Island*

*Melville
Island*

GREENLAND

*Banks
Island*

*BAFFIN
BAY*

Mackenzie R.

ALASKA

*Victoria
Island*

Baffin Bay

Baffin Island

Fort Good Hope

Fort Franklin

Fort Norman

Coppermine R.

Back R.

*FOXE
BASIN*

Fort Enterprise

Fort Providence

Fort Resolution

*Southampton
Island*

*HUDSON
BAY*

Fort Chipewyan

*Dubwant
Lake*

Fort Churchill

Cumberland
House

York
Factory

Prince Albert

Norway House

0	500 kms

CANADA

0	300 mls

USA

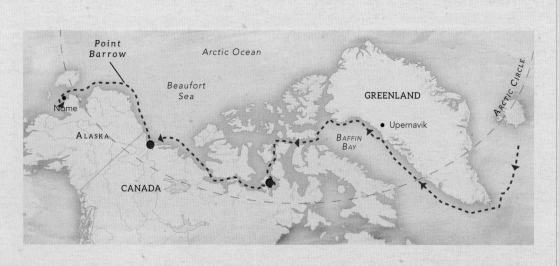

Point
Barrow

Arctic Ocean

*Beaufort
Sea*

GREENLAND

ARCTIC CIRCLE

Nome

ALASKA

*BAFFIN
BAY*

Upernavik

CANADA

◄━━━	John Ross, 1818
◄┄┄┄	John Ross, 1829–1833
◄━━━	Parry, 1819–1820
◄┅┅┅	Parry, 1821–1823
◄━━━	Back, 1833–1834
◄━━━	Hearne, 1770–1772
◄━━━	Mackenzie, 1789
◄━━━	Franklin, 1st expedition
◄┅┅┅	Franklin, 2nd expedition
◄┅┅┅	Franklin, 2nd expedition – Richardson's route
◄┅┅┅	Franklin, last expedition
◄┅┅┅	Amundsen, 1903–1906
●	Amundsen's winter camps

THE NORTHWEST PASSAGE

From the time of Martin Frobisher and John Davis, the goal of finding a Northwest Passage always held a special place in the British psyche. In 1744, Parliament offered £20,000 to the first person to complete the Passage. Beginning in 1769, Samuel Hearne of the Hudson's Bay Company made three overland journeys through northern Canada to investigate rumoured copper deposits and the existence of a low-latitude Northwest Passage. On his final trek, Hearne followed the Coppermine River to the Arctic Ocean, becoming the first European to stand on the north coast of the Americas. Then, in 1789, Alexander Mackenzie followed a river north from Great Slave Lake. Now named the Mackenzie River, it took him not to the Pacific as he hoped, but to the Arctic Ocean, making it a second known point (with the mouth of the Coppermine) along that otherwise mysterious coast.

BELOW LEFT John Ross, shown here, was not fortunate on his three Arctic expeditions. Once he believed his passage was blocked by non-existent land; later his ship was trapped in the ice and abandoned; and finally, he left his personal yacht behind.

BELOW MIDDLE William Edward Parry was the beau ideal of Arctic explorers. A skilled navigator, exceptional leader, and dashing public figure, he was knighted, rose to admiral, and was a key player in the higher echelons of the Admiralty.

BELOW RIGHT A portrait of John Franklin based on his first expedition. He is shown seated among cartographic and navigational instruments, while in the background is Fort Enterprise, one of his key bases.

The end of the Napoleonic Wars left the Royal Navy with officers, men, and ships unemployed. At the urging of John Barrow, Second Secretary of the Admiralty, many were reassigned to exploring expeditions, including, in 1818, two voyages to the far north. One, under David Buchan, was the Royal Navy's last – and unsuccessful – attempt to sail to the North Pole. The other, to search for a Northwest Passage, was commanded by John Ross. After being stopped by ice in north Baffin Bay, Ross entered Lancaster Sound, but soon turned back, claiming it was blocked by land. His second-in-command,

William Edward Parry, disagreed, but the expedition nevertheless returned to Britain.

Ross was soundly criticized, and the next year Parry was ordered to re-examine Lancaster Sound. He sailed right through the supposed mountains, continued along Barrow Strait, and discovered numerous islands. Stopped by ice at Melville Island, Parry's party became the first to winter in the high Arctic intentionally. After exploring on foot, and proceeding as far as the ice would allow in the spring, Parry returned to England, having discovered the entrance to the Northwest Passage. Parry led two more attempts, including an examination of the coast of the Foxe Basin in 1821–23, finding in the process another entrance, although it proved impassable to sailing ships.

Concurrently with Parry's initial voyage, John Franklin, who had commanded the second ship under Buchan, was sent to explore the northern coast of America. With Dr John Richardson, two midshipmen, and a party of Canadian voyageurs – French-Canadian or part-Indian trappers and traders hired to guide and help transport supplies – he made slow progress across the Canadian river systems and to the Arctic coast, but conducted detailed surveys and compiled scientific collections. When the party ran out of food on the return,

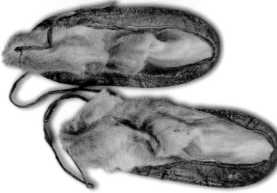

TOP This image, from John Ross' expedition in *Isabella and Alexander* in 1818, is typical of its time. Dangers tended to be exaggerated, with icebergs larger and sharper than usual and polar bears bigger, more vicious and undoubtedly hungrier.

LEFT Among the fascinating items held by the Royal Geographical Society are these Inuit boots brought back by William Edward Parry in 1823 at the end of his second Arctic command.

NORDENSKIÖLD AND THE NORTHEAST PASSAGE

Adolf Erik Nordenskiöld led six Swedish scientific, exploratory, and commercial expeditions to the Arctic before, with sponsorship from King Oscar II and merchants Oscar Dickson and Aleksandr Sibiryakov, completing the first navigation of the Northeast Passage. Sailing in *Vega* in June 1878, Nordenskiöld made a quick transit along most of the Siberian coast and passed Cape Chelyuskin, the northernmost point of Asia. Stopped by ice 200 kilometres (120 miles) west of Bering Strait, Nordenskiöld put into winter quarters until mid-1879. He was then easily able to complete the Passage and sailed south through the Bering Strait.fulfilled the goal of centuries.

many died, and others were forced to eat lichen and leather clothes and boots. Midshipman Robert Hood was murdered by a voyageur, who was then executed. In 1822, Franklin finally struggled home, where he became a popular hero.

Three years later, Franklin returned to extend his discoveries. He and Richardson ascended the Mackenzie River and then went in separate directions, between them mapping approximately half the North American Arctic coast. Ross also returned in 1829, commanding a private venture in a paddle steamer. Proceeding down Prince Regent Inlet, the ship was caught in the ice and was only able to move slightly in the next two years. During this time James Clark Ross, second-in-command and the nephew of John Ross, made extensive sledging trips, including in 1831 becoming the first man to attain the North Magnetic Pole. After abandoning the ice-bound ship, the party wintered on Somerset Island, then took boats down Lancaster Sound, where in 1833 they were rescued by whalers.

Meanwhile, George Back, who had served on both of Franklin's expeditions, descended the Great Fish River and explored parts of the coast while making an unsuccessful search for Ross. A decade later, in 1845, the 59-year-old Franklin set off with 129 men and two ships, *Erebus* and *Terror*, on the most lavishly appointed Northwest Passage expedition ever mounted. For several years, no news came, and then people nervously began to realize that the entire expedition had vanished.

THE NAVIGATION OF THE NORTHWEST PASSAGE

After so many British attempts, it was a Norwegian, Roald Amundsen, who finally navigated the Northwest Passage. Born in 1872, Amundsen grew up dreaming of succeeding where his hero John Franklin had not. He earned a master's sailing license, made himself an expert skier and served as mate on *Belgica*, the first ship to winter in the Antarctic. In 1903, he took the 47-ton sloop *Gjøa* and six men into Lancaster Sound. They spent three winters *en route*, but in 1906 completed the Passage, having fulfilled the goal of centuries.

OPPOSITE This famous paingting shows the August 1818 initial encounter of John Ross and William Edward Parry with people whom Ross called "Arctic Highlanders". The term did not stick, and they soon bcme known as "Polar Eskimos".

ABOVE AND LEFT Roald Amundsen, conqueror of the Northwest Passage, and some of the Inuit amongst whom he spent three winters while accomplishing the feat.

C. Bird
BELLOT STRAIT
BRENTFORD BAY
Grimble I.
CAPT. M'CLINTOCK ARRIVED 20TH AUG DEPARTED 9TH AU

PASSED THROUGH BY THE YACHT FOX 6TH SEPT & REMAINED TILL 27TH SEPT. TIDES 6 TO 7 KNOTS IN THE SPRINGS

C. Hodgkin
Port Kennedy
Murray B.
C. Scoresby

1ST JOURNEY 1859
DEPARTED 17TH FEB.
RETURNED 13TH MARCH
2ND JOURNEY
DEPARTED 2ND APRIL 1859
RETURN M'CLINTOCK 19TH JUNE
HOBSON 14TH
YOUNG 28TH

Babbage Bay
C. Airy
Port Logan
Moltke B.

Pt Allen Young

Aoland Bay

Cape Swinburne

M'CLINTOCK CHANNEL

SUPPOSED ROUTE OF THE EREBUS & TERROR

EXPLORED BY M'CLINTOCK & HOBSON

BOOTHIA FELIX

J. Owen

M'CLINTOCK & HOBSON PARTED HERE ON FINAL SEARCH 28TH APRIL

28TH FEB. 1859 CAPT. M'CLINTOCK FOUND NATIVES & RELICS.

Gateshead I.
Collinson's farthest 1852
C. Collinson

C. Nikolai
Larsen B.
Ross' farthest 1831

SIR JOHN ROSS WINTERED 1829 30 31

Krusenstern Lakes

BOOTHIA ISTHMUS

EREBUS & TERROR BESET IN THE ICE 12TH SEPT. 1846

Pelly Pt
Rae's farthest 1851

C. Adelaide Regina
Magnetic Pole

Kent B.

L. Hansteen
L. Jekyll

EREBUS & TERROR ABANDONED 22ND APRIL 1848

Halkett I.

C. Felix
Walls B.
C. Maria Louisa

Beaufort on Clarence
C. Sidney

Kent B.
C. Victoria
C. Gloucester
Oscar B.

C. Sussex

C. Maria de Gloria
R. Garry

ROSS CAIRN REACHED BY HOBSON 6TH MAY M'CLINTOCK 5TH JUNE

CAIRNS WITH TENTS BLANKETS

CAPT. CROZIER LANDED HERE IN COMMAND OF 105 MEN THE REMAINS OF THE CREW OF THE EREBUS & TERROR AFTER ABANDONING THE VESSELS ON 22ND APRIL 1848. STARTED FOR GREAT FISH RIVER 26TH APRIL 1848.

C. Alfred

Victory Pt.
Ross' farthest
Back B.

CAIRN WITH BROKEN PICKAXE & CANISTER
ROSS CAIRN RECORD FOUND

C. Sophia
C. Sabine
Beverley Is.

Christian Frederick

Spence B.

Albert Edward Bay
C. Adelaide
Admiralty I.

C. Franklin

Port Parry

Matty Id.

Catherine I.
Willersted L.

Taylor I.
Drift Wood Pt

CAIRN WITH RECORD DEPOSITED BY LT GORE & M. DES VOEUX MAY 1847

KING WILLIAMS ISLAND

Edgeworth
C. Norton
C. Smyth
Ross 1830

C. Porter
Ross 1830

Balfour B. Rae 1854

8TH MAY SNOW VILLAGE WITH NATIVES & RELICS

BOAT'S MAST FOUND BY D'R RAE 21ST AUG 1851
Parker B.

LOW SHINGLE BANKS & LOW ISLES
NATIVES SEEN

BOAT FOUND 28 ft. LONG 7 ft. 3 in. WIDE, 2 ft. 4 in. DEEP CONTAINING 2 HUMAN SKELETONS, CLOTHING, 2 GUNS, 5 WATCHES, SPOONS, FORKS, &c.

Dehaven Pt
Erebus B.
C. Crozier

VICTORIA STRAIT

Terror B.
SUPPOSED POSITION OF WRECK

NO TRACE OF WRECK OR NATIVES SEEN

Herschell
Low Limestone Tract

FOUND 24TH MAY A BLEACHED HUMAN SKELETON WITH POCKET BOOK AND LETTERS, FRAGMENTS OF CLOTHING &c.

De la Guiche Pt.

C. Colville
Acland Pt.

Jenny Lind I.

Monument & wreck here July 17 1859
Dease and Simpson
Gladman Pt
Macgillivray B.
C. Geddes

GREAT BAY

Senforth
Grant Pt
Reid I.
Wilmot

Pt Jas. Ross

Todd Is.
Douglas B.
Pfeiffer
Halloch Pt
Richardson Pt
Macnochie I.

Inglis B.
C. Selkirk

ADELAIDE PENINSULA

Pt. Booth Back 1834 SNOW HUT WITH NATIVES & RELICS

White Bear Pt
Gernon B.

Thunder Cove
O'Reilly I.

Pt. Sir Chas. Ogle TRACES FOUND 1856 BY MR ANDERSON

Castor & Pollux Rae Dease & Simpson 1834 1839
OBTAINS SOME RELICS FROM ESQUIM

Keith Pt
Johnson Pt
Macloughlin B.
Bowes Pt

Barrow Inlet

Ripon Id.
Pt Britannia
Mount I.

TRACES FOUND 1855

C. Hay
Beaufort

Blackwood Pt
Chester B.
Stewart Pt

Pt. Peckell

Ogden B.

Elliott Bay

Pt. Gage
Victoria Headland
Backhouse Pt.

Cockburn B.

SKETCH
of the
RECENT DISCOVERIES
on the
NORTHERN COAST of AMERICA
BY
CAPTAIN M'CLINTOCK R.N.
in Search of
SIR JOHN FRANKLIN.

Lake Franklin

Whirlpool Rapid
Montresor Rapid

BACK TRACES FOUND HERE BY MR ANDERSON 1855

London, Published by Jas. Wyld, Geographer to the Queen.

Charing Cross East

PLACES WHERE RELICS OF THE FRANKLIN EXPEDITION HAVE BEEN FOUND ...... ○ CAIRN

COAST EXPLORED BY THE SEARCHING EXPEDITIONS

PROBABLE ROUTE OF THE FRANKLIN EXPEDITION

THE FRANKLIN SEARCHES

By mid-1847, the Admiralty had received no word from John Franklin's expedition, so it sent his old colleague John Richardson to explore the coasts between the Mackenzie and Coppermine rivers. The next year, it despatched a ship-based search under James Clark Ross, who, coincidentally, had turned down command of the Northwest Passage expedition, allowing Franklin to step in. Neither found any trace of Franklin's men or ships. Finding Franklin's expedition soon became a consuming passion for the British public and Admiralty alike, and over the next dozen years, more than 30 missions sought traces of it, in the process solving many of the geographical mysteries surrounding the Canadian archipelago.

OPPOSITE Following the disappearance of Sir John Franklin's 1845 expedition in search of the Northwest Passage, there were several attempts to find it. Captain Francis Leopold McClintock, an Irish explorer in the British Royal Navy, joined a series of searches for Sir John Franklin between 1848 and 1859. In 1857 he commanded the yacht Fox, which was sponsored by public subscription via Lady Jane Franklin to search for her missing husband. On this trip, in May 1859, he found the only official record of the 1845–8 Sir John Franklin Northwest Passage Expedition. McClintock's account of his trip was published in The Voyage of the Fox in 1859.

RIGHT Francis Leopold McClintock was one of the most successful exponents of the British man-hauling tradition. His greatest fame came after his expedition that "solved" the mysteries of Franklin's disappearance.

Expeditions searched west from Lancaster Sound, east from Bering Strait, north from the Canadian mainland and in the waterways north of where the Passage was supposed to lie. They were sent by the Royal Navy, sponsored by Lady Franklin or public subscription, carried out as addenda to whaling voyages, subsidized by the Hudson's Bay Company and even financed by the American merchant Henry Grinnell.

The winter of 1850–51 saw a glut of expeditions in Lancaster Sound. Horatio Austin, commanding a squadron of four ships, found Franklin's wintering site of 1845–46 at Beechey Island and then sent out numerous sledging trips, exploring new areas, at the same time laying the foundations of the Royal Navy man-hauling tradition. Austin was joined at Beechey by expeditions under John Ross, the whaler William Penny and the American commander of Grinnell's expedition, Edwin Jesse De Haven, none of which recorded any great success. In 1852, the four ships

that had been under Austin's command sailed once more under Edward Belcher. As before, extensive sledging was conducted, with Francis Leopold McClintock proving exceptionally adept at it.

Meanwhile, in 1850, the Admiralty sent two ships to search for Franklin via the Bering Strait. In HMS *Enterprise*, Richard Collinson reached a position not far from Franklin's farthest point west, but after being forced to winter three times retreated to Britain. Robert McClure in HMS *Investigator* also spent three harsh winters in the western Arctic. In 1853, his party facing rampant scurvy, he tried to reach safety by walking east over the ice. The group did not go far before being met by Belcher's expedition. Eventually, McClure sailed for England on a supply vessel after Belcher ordered four of his ships to be abandoned. McClure and his men had made the first completion of the Northwest Passage, although part had been on foot.

Belcher and McClure were not the only Franklin search parties to abandon ship. In 1853, Elisha Kent Kane, the surgeon under De Haven, led Grinnell's second expedition. Although he learnt little new about Franklin, before abandoning ship, Kane discovered Kane Basin, north of Smith Sound, which

would become a key route towards the North Pole. Much more successful was John Rae of the Hudson's Bay Company, who in 1854 obtained relics from Inuit and statements that the bodies of many white men had been found on King William Island, where they had died of starvation and scurvy. The Admiralty and the British public grudgingly accepted Rae's evidence, but many were inflamed by his conclusion that the party had engaged in cannibalism.

ABOVE A sledging party leaves HMS *Investigator* during McClure's four-year expedition in the Canadian Arctic. In 1853 the ship was abandoned and the crew transferred to HMS *Resolute*, which in turn became beset in Barrow Strait.

LEFT Robert McClure portrayed by Stephen Pearce. McClure was widely acclaimed as the man who completed the Northwest Passage, despite part of it being on foot. Lady Franklin insisted that honour belonged to her husband.

OPPOSITE A map used by Francis Leopold McClintock on his last Franklin search expedition. He had copied it from an Inuit map on a previous expedition.

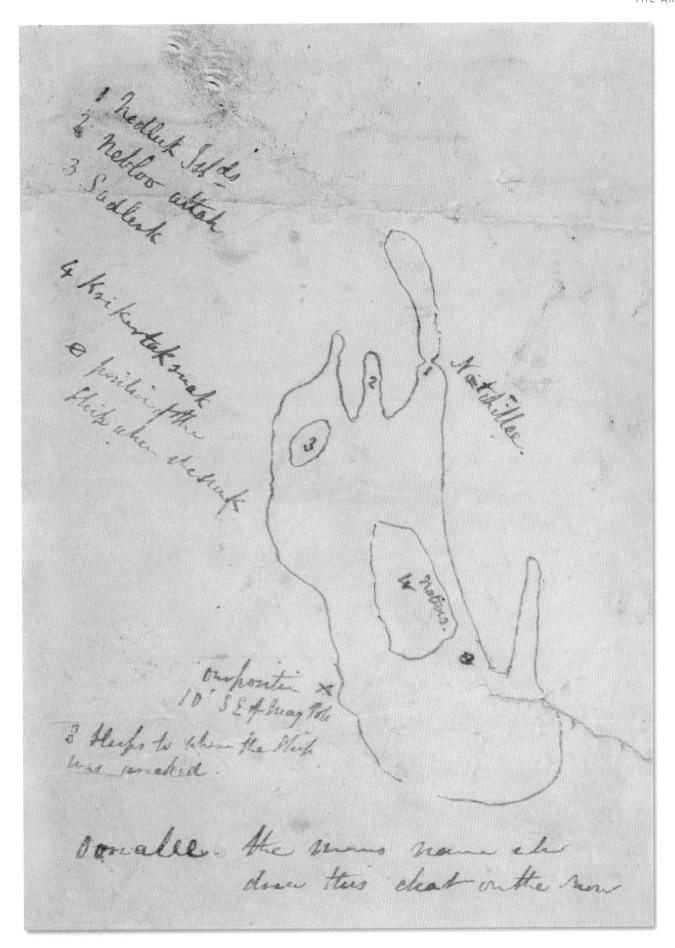

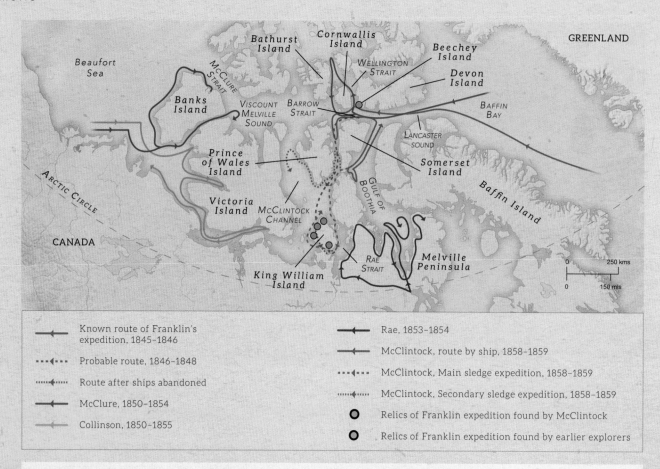

——●—— Known route of Franklin's expedition, 1845–1846	——◄—— Rae, 1853–1854
·—·◄·—· Probable route, 1846–1848	——◄—— McClintock, route by ship, 1858–1859
·····◄····· Route after ships abandoned	·—·◄·—· McClintock, Main sledge expedition, 1858–1859
——◄—— McClure, 1850–1854	·····◄····· McClintock, Secondary sledge expedition, 1858–1859
——◄—— Collinson, 1850–1855	◯ Relics of Franklin expedition found by McClintock
	◉ Relics of Franklin expedition found by earlier explorers

MAJOR FRANKLIN SEARCHES, 1847–59

YEARS	COMMANDER	SHIPS	YEARS	COMMANDER	SHIPS
1847	William Penny	St Andrew	1850–51	John Ross	Felix & Mary
1847–49	John Richardson & John Rae	overland	1850–51	Edwin Jesse	Advance & Rescue De Haven
1848–49	James Clark Ross	HMS Enterprise Investigator	1850–51	John Rae	overland
			1850–51	John James Barnard	overland
1848–50	Henry Kellett	HMS Herald	1850–54	Robert McClure	HMS Investigator
1848–52	Thomas Moore	HMS Plover	1850–55	Richard Collinson	HMS Enterprise
1849	William Penny	Advice	1851–52	William Kennedy	Prince Albert
1849	Robert Shedden	Nancy Dawson	1852	Edward Inglefield	Isabel
1849–50	James Saunders	HMS North Star	1852–54	Edward Belcher	HMS Assistance & Pioneer
1849–51	William Pullen	boats			
1850	Charles C. Forsyth	Prince Albert	1852–54	Henry Kellett	HMS Resolute & Intrepid
1850–51	Horatio Austin	HMS Resolute, Assistance, Intrepid & Pioneer	1852–54	William Pullen	HMS North Star
			1852–54	Rochfort Maguire	HMS Plover
			1853–54	John Rae	overland
1850–51	William Penny	HMS Lady Franklin & Sophia	1853–55	Elisha Kent Kane	Advance
			1855	James Anderson & James Stewart	overland
			1857–59	Francis Leopold McClintock	Fox

ABOVE The graves of three of Franklin's men – John Torrington, John Hartnell, and William Braine – who died at Beechey Island in 1846. On the far right is the grave of Thomas Morgan of Investigator, who died in 1854.

BELOW The body of Petty Officer John Torrington, which in 1984 was exhumed briefly at Beechey Island along with John Hartnell and William Braine. They were found to be remarkably well preserved by the permafrost.

Unwilling to accept Rae's assessment, in 1857 Lady Franklin sponsored an expedition led by McClintock. On King William Island, McClintock obtained numerous relics from Franklin's expedition, and William Hobson found a written record that told of Franklin's death in 1847, of how the ships were frozen in the ice for more than a year before being abandoned in 1848, and of how the men had tried to make their way to safety. With McClintock's return to England, the Franklin mystery had at last been solved, although later expeditions continued to find relics and bodies, and showed that the last survivors had reached Adelaide Peninsula before dying. As it turned out, Franklin's triumph had not been the Northwest Passage, but rather that the search for him had investigated new areas that might otherwise have taken decades to explore.

LADY FRANKLIN

Jane Franklin (c.1792–1875), née Griffin, married John Franklin in 1828. An energetic and forceful woman, she helped influence her husband's progressive policies as Lieutenant Governor of Van Diemen's Land. Following the disappearance of his last expedition, she kept constant pressure on the Admiralty to maintain their search. She funded several expeditions herself or with the help of public subscription, including McClintock's. In 1857, she became the first woman to receive the Gold Medal of the RGS, in recognition of her role in the exploration of the Canadian archipelago.

RIGHT A bust of Lady Franklin, held at the Royal Geographical Society. Her continual efforts to discover the fate of her husband and his expedition were finally successful under Francis Leopold McClintock and William Hobson.

FRIDTJOF NANSEN

In 1884, relics from *Jeannette* (see page 170), which had been crushed near the New Siberian Islands, were found on southwest Greenland. This find was insignificant to most people, but to Norwegian explorer and scientist Fridtjof Nansen it indicated the existence of a trans-polar current in the Arctic Basin. To prove his theory, Nansen proposed that a specially built ship be deliberately beset in the ice so the current could carry it clear across the Arctic Ocean. The scientific community was highly sceptical and proclaimed Nansen reckless and foolhardy. But he was neither: he was the greatest intellect and most creative mind in the history of polar exploration.

OPPOSITE TOP Nansen (second left) and Hjalmar Johansen (right) before their departure from *Fram* in March 1895. Otto Sverdrup and four others travelled with them for part of the first day because the explorers had experienced problems in their earlier attempts to leave.

OPPOSITE BOTTOM Nansen working with one sledge dog. Nansen was an early advocate of using dogs to pull sledges while skiing next to them. Many of the principles for this method were further worked out by Otto Sverdrup.

TOP Nansen reading the temperature. Once *Fram* was frozen into the ice, a number of small scientific stations were set up in the area around her to assist in meteorological, biological, oceanographic, and ice studies.

Born in 1861, Nansen first visited the Arctic in 1882 to improve his zoological knowledge. He thereafter studied numerous aspects of Arctic travel while also producing a pioneering doctoral thesis. In 1888–89, he announced his presence on the international Arctic scene with a flourish, leading the first expedition to cross the Greenland icecap.

Many explorers had unsuccessfully attempted to cross Greenland, but Nansen's innovative plan set him apart. Instead of starting from the inhabited west coast, his party was dropped near the sparsely populated east coast. This not only meant that it had to cross the icecap only once, but that there was no turning back: the supplies needed for salvation were on the west

coast. The six men made slow progress at first, but once on the icy plateau, they virtually sped to the west coast, where Nansen and Otto Sverdrup built a small boat and rowed to the capital, Godthåb. On the expedition, Nansen had not only demonstrated that the icecap extends across Greenland, but showed the value of skis for polar exploration, proving they functioned in a wide variety of snow conditions. He also designed the light, flexible Nansen sledge and the innovative Nansen cooker, both of which remained standard for polar explorers for decades.

This, then, was the man who continued with his audacious plan, and September 1893 saw his new ship *Fram* locked into the ice. Within the next year,

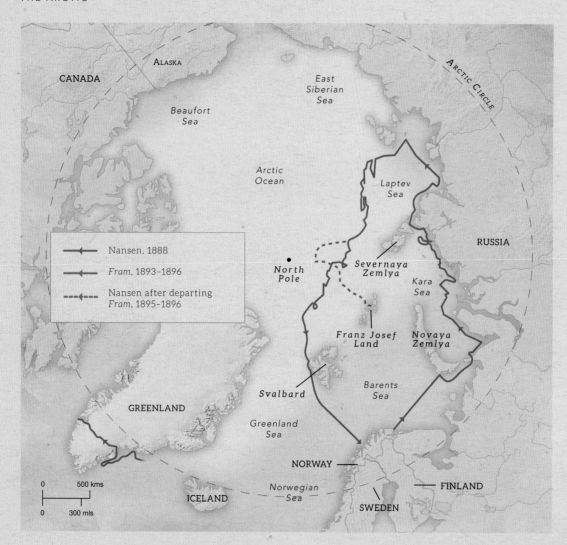

Nansen, 1888
Fram, 1893–1896
Nansen after departing
Fram, 1895–1896

BELOW Otto Sverdrup in his cabin aboard *Fram*. His four-year expedition (1898–1902) explored an entirely new region of the Canadian Arctic, discovering Axel Heiberg, Amund Ringnes and Ellef Ringnes islands.

measurements made aboard confirmed the existence of the current and proved the Arctic Basin to be a deep ocean. But it became obvious that the drift, although taking them towards Greenland, would bypass the North Pole. So Nansen prepared new plans: he would leave the moving ship, despite its being in an uncharted area of the Arctic Ocean, and knowing that he had no chance of finding it again. Then he would sledge to the Pole, and return over unknown ice conditions to either Franz Josef Land or Svalbard, both uninhabited and many weeks travel away, and he would do it with only one companion, 28 dogs and two kayaks.

In the middle of March 1895, in latitude 84° 4' N, Nansen left the ship under the command of Sverdrup and headed north with Hjalmar Johansen. They reached 86° 13' N, 275 kilometres (170 miles) closer to the Pole than ever attained before. Then they retreated 800 kilometres (500 miles) to Franz Josef Land, where they built a tiny hut and spent the

OTTO SVERDRUP

Although little remembered, Otto Sverdrup (1854–1930) was one of the most accomplished polar explorers. He was Nansen's closest companion on the crossing of Greenland, and then captained *Fram* on the polar drift. Between 1898 and 1902, Sverdrup led the second *Fram* expedition, discovering major islands in the Canadian archipelago and carefully exploring and charting some 260,000 km2 (160,000 square miles), the largest area ever on an Arctic expedition. He also perfected the technique of driving a sledge while skiing next to it. He later led several minor Arctic expeditions.

winter living on walrus and polar bear. In spring they headed southwest before an unlikely meeting with the English explorer Frederick George Jackson, who had been exploring in Franz Josef Land for almost two years. Meanwhile, *Fram* continued drifting, attaining her farthest north at 85° 55' N, and finally reaching Tromsø on 20 August 1896, one day before Jackson's ship arrived with Nansen and Johansen.

Nansen's unexpected return made him the darling of the international press, a role for which he was perfectly suited, being tall, photogenic, intellectual and graceful. Although he did not lead another major expedition, he became the great mentor whom all others consulted in polar matters. He also was named professor of oceanography at the University of Christiania (the city was renamed Oslo in 1925), served as Minister to the Court of St James when Norway gained independence, and received the Nobel Peace Prize for his work repatriating prisoners of war after the First World War. He died in 1930.

FRAM

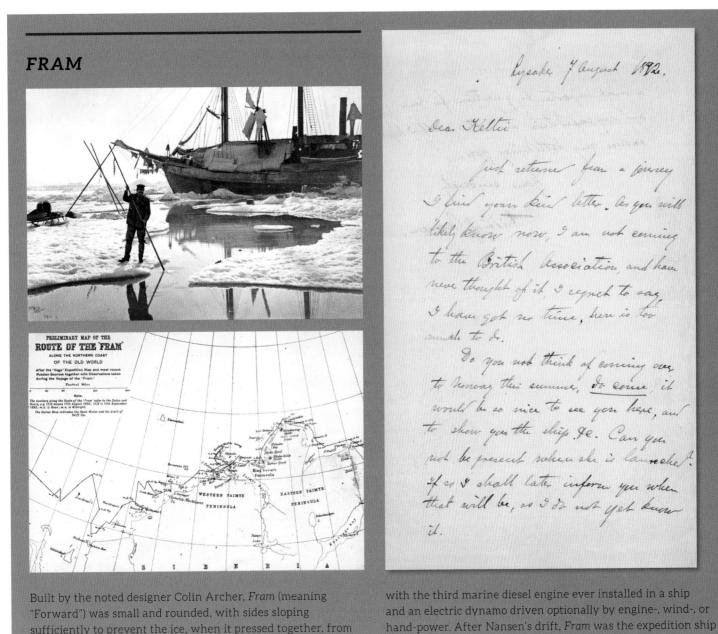

Built by the noted designer Colin Archer, *Fram* (meaning "Forward") was small and rounded, with sides sloping sufficiently to prevent the ice, when it pressed together, from getting a hold on the hull. Thus, rather than crushing her, the ice simply raised her out of the water. *Fram* was also equipped with the third marine diesel engine ever installed in a ship and an electric dynamo driven optionally by engine-, wind-, or hand-power. After Nansen's drift, *Fram* was the expedition ship for Sverdrup and then for Amundsen's South Polar expedition (1910–12).

ABOVE In September 1875 HMS *Alert* reached 82°28'N, the northernmost latitude yet reached by a ship, before putting into winter quarters. The next April, Markham set out from there on his record-breaking sledge journey.

LEFT The three open boats from *Jeannette* were tragically separated in the middle of a storm as they attempted to reach the Lena Delta.

THE NORTH POLE

Throughout the Royal Navy's intense hunt for the Northwest Passage, and the subsequent Franklin searches, there was no great emphasis on reaching the North Pole, although William Edward Parry did make one attempt in 1827, using boats fitted with sledge runners for travel over both ice and water. Parry struggled north from Svalbard for a month before realizing the drift of the ice was taking his party south almost as quickly as they were going north. He abandoned the effort after reaching a record north of 82° 45' N.

ABOVE The terrible man-hauling sledge trip undertaken by Albert Hastings Markham and his men was a far cry from the ease and comfort experienced by the officers of HMS *Alert and Discovery* in their comfortable wardrooms.

Robert McClure's completion of the Passage (although partly on foot) and Francis Leopold McClintock's discovery of the Franklin party's fate left the North Pole as the next goal that gripped both explorers and the public. In 1860–61, Isaac Israel Hayes, previously the surgeon on Elisha Kent Kane's expedition, attempted to reach the Pole via Kane Basin, between Greenland and Ellesmere Island. He was expecting to find an "open polar sea" – an ice-free body of water around the Pole encircled by a ring of ice – but was stopped short of the north coast of Greenland. A decade later, Hayes's route was followed by another American, Charles Francis Hall, who had attracted attention worldwide for his two lengthy searches for Franklin relics, during which he had adopted Inuit methods of travel and living. Now commanding an

official American expedition, Hall sailed to north Greenland, but died mysteriously in November 1871. The following spring, his men retreated south before being split in two; both parties ultimately required rescuing.

The next major attempt was the Admiralty's British Arctic Expedition of 1875–76 under George Strong Nares, who also followed the "American route" between Greenland and Ellesmere. In spring 1876, a man-hauling sledging party led by Albert H. Markham attained a record latitude of 83° 20' 26" N, but were driven back by scurvy. Widespread scurvy throughout the crew then forced Nares to return to England a year earlier than planned.

By this time, the world's foremost proponent of the notion of an open polar sea was the German

AUSTRO-HUNGARIAN EXPLORING EXPEDITION

The *Jeannette* expedition was not the first to test Petermann's concepts. In 1869, Karl Koldewey led a German attempt to find a route north via east Greenland. Then in 1872, Petermann's theories helped launch the Austro-Hungarian Exploring Expedition, an effort to investigate unknown Arctic regions under the joint command of Karl Weyprecht and Julius Payer. However, their ship *Tegetthoff* was caught in the ice and drifted helplessly northwest. In August 1873, they discovered the archipelago of Franz Josef Land; the next year, they abandoned ship and reached Novaya Zemlya in four open boats.

LEFT In a painting by Julius Payer, his joint commander, Karl Weyprecht persuades the frightened members of the crew that they cannot return to *Tegetthoff* after abandoning their expedition ship.

geographer August Petermann. In 1879, his theory led James Gordon Bennett, owner of *The New York Herald*, to send an expedition under George Washington De Long through the Bering Strait towards the Pole. However, the ship *Jeannette* was soon trapped in the ice north of Siberia, where it drifted for more than a year before being crushed. Three boats headed towards the Lena Delta, but although the group under George Melville eventually reached safety, one boat was never seen again, and De Long's party died of starvation and exposure in the Delta.

Another American tragedy soon followed. In 1881, Adolphus W. Greely led a US expedition to northern Ellesmere Island as part of the International Polar Year co-operative scientific effort. The following spring, James Lockwood sledged to 83° 24' N, breaking Markham's record. But the relief ship never reached Greely's base, and after another winter he led his men south by boat and foot. They were forced to winter on southern Ellesmere, and 18 men died, mostly of famine and scurvy, although one committed suicide and another was executed. Greely and six others were finally rescued in June 1884, following which the expedition received international attention due to evidence of cannibalism.

The American farthest north stood until 1895, when Fridtjof Nansen reached 86° 13' 06" N. Amazingly, it was only four years later that this was surpassed on an expedition led by Luigi Amedeo di Savoia, Duke of the Abruzzi. From northern Franz Josef Land, an Italian party of 10 men and 102 dogs under Umberto Cagni set out, with the final detachment reaching 86° 34' N in April 1900. Unfortunately, one of the three-man support parties disappeared on its return and was never found.

Numerous less successful attempts on the Pole were made near the turn of the century from Franz Josef Land, including by Frederick George Jackson (1894–97), Walter Wellman (1898–99), Evelyn Baldwin (1901–02), and Anthony Fiala (1903–05). Wellman also made four attempts from near Spitsbergen, the last three by airship.

OPPOSITE TOP Lieutenant George Washington De Long on the eve of his departure from San Francisco. His tendency to be a martinet did not make for a happy ship.

OPPOSITE BOTTOM Adolphus W. Greely in old age, after he had been promoted to general; he was widely considered one of the foremost Arctic experts in the United States.

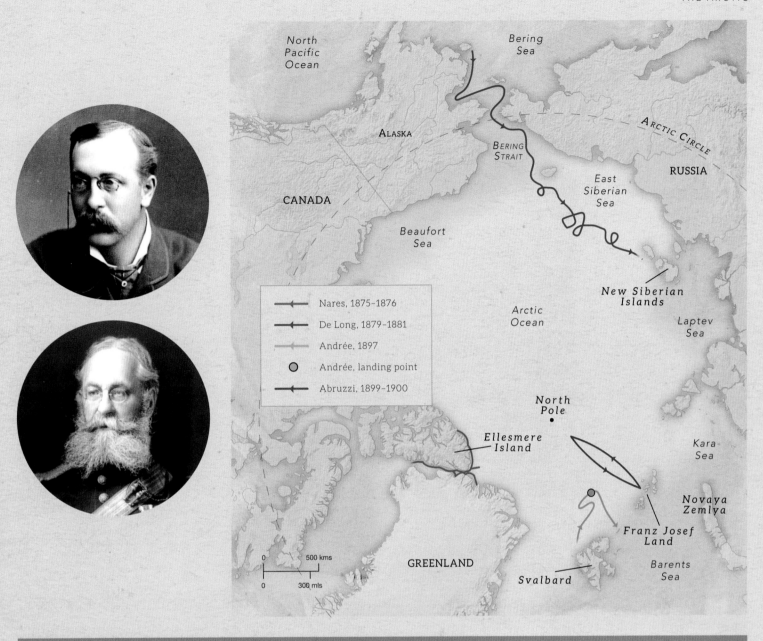

Legend:
- Nares, 1875–1876
- De Long, 1879–1881
- Andrée, 1897
- ⦿ Andrée, landing point
- Abruzzi, 1899–1900

ANDRÉE'S FLIGHT

In 1897, Salomon August Andrée, a Swedish physicist attempted to reach the North Pole using a hydrogen-filled balloon named *Örnen*. He took off from Danskøya, near Spitsbergen, with two companions and disappeared. Their fate remained a mystery until 1930, when a Norwegian expedition discovered their bodies, diaries and film, which proved the balloon had landed on the ice some 325 kilometres (200 miles) northeast of Svalbard, leaving the three obliged to walk to safety. They reached the little-known island of Kvitøya, where they died of carbon monoxide asphyxiation, food poisoning or hypothermia.

RIGHT Andrée's balloon suffered damage at take-off and could not be adequately controlled. Sixty-five hours later, after touching down several times, he and his companions, Nils Strindberg and Knut Fraenkel, landed on the ice once and for all.

THE NORTH POLE CLAIMED

The history of exploration is marked by many questions and many debates, few of them longer lasting and none more vehement than that of who was actually first to reach the North Pole. It began in September 1909, when two men almost simultaneously claimed the honour.

OPPOSITE TOP Peary's final five companions on his last expedition, standing at, he claimed, the North Pole. Matthew Henson, his long-time assistant is in the centre under the flag.

OPPOSITE BOTTOM LEFT The tri-motor Fokker *Josephine Ford*, which was piloted by Floyd Bennett when Byrd claimed they had flown to the North Pole and back in under 16 hours.

OPPOSITE BOTTOM RIGHT Peary had this picture specially photographed and tinted to give just the impression of him that he wanted. In his later expeditions – after losing his toes – it would have been rare for him to travel on foot like this.

TOP The cover of the French publication *Le Petit Journal* for 19 September 1909, highlighted the conflicting claims of Robert E. Peary and Frederick Cook by showing them fighting over the North Pole. Notice the penguins, which the artist did not realize were not found in the Arctic.

One was already an American hero: Robert E. Peary had first gone to Greenland in 1886, and in the next dozen years he led another four expeditions there. Sponsored by the Peary Arctic Club, a group of wealthy and patriotic businessmen, he then made two attempts to reach the North Pole. On the first – in 1898–1902 – he lost seven toes due to frostbite from an ill-advised mid-winter journey, and on the second he claimed to have reached a record latitude of 87° 06' N. The other man, Dr Frederick Cook, had first gone to the Arctic under Peary in 1891–92, had played a major role in the survival of the crew of *Belgica* during the first Antarctic wintering (1898), and had claimed the first ascent of Mount McKinley, the highest mountain in North America in 1906.

In 1908, Peary left on his third and last North Polar expedition. When he reached Labrador on his return the following year, he discovered that in that very week Cook had announced that he had reached the North Pole. Peary immediately claimed to have attained it himself, and denounced Cook

as a charlatan. For months, controversy swirled, before most newspapers and geographical societies hesitantly recognized Peary's claim and rejected Cook's. But that was not the end of the matter, as doubts remained about Peary's supposed distances travelled, his diary and his behaviour before and after "reaching the Pole". Today, most unbiased polar historians agree that neither man reached the Pole.

Remarkably, the next claim to have reached the North Pole was another hoax. In 1925, Roald Amundsen, the conqueror of the Northwest Passage and the South Pole, tried to fly to the North Pole with the American aviator Lincoln Ellsworth but was forced to land short of the Pole. The next year, they were back with a dirigible, designed and piloted by the Italian Umberto Nobile and named *Norge*. Before they took off from Spitsbergen, however, an American naval officer, Richard E. Byrd, arrived with his own hopes of making the first flight to the Pole. Wishing to avoid another "race to the Pole", Amundsen calmly waited while Byrd prepared his

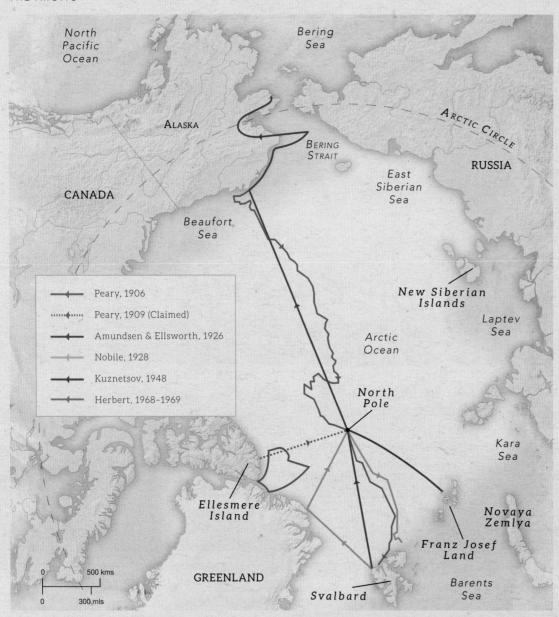

Peary, 1906
Peary, 1909 (Claimed)
Amundsen & Ellsworth, 1926
Nobile, 1928
Kuznetsov, 1948
Herbert, 1968–1969

North Pacific Ocean

Bering Sea

ARCTIC CIRCLE

ALASKA

BERING STRAIT

RUSSIA

East Siberian Sea

CANADA

Beaufort Sea

New Siberian Islands

Laptev Sea

Arctic Ocean

North Pole

Kara Sea

Novaya Zemlya

Ellesmere Island

Franz Josef Land

Barents Sea

GREENLAND

Svalbard

0   500 kms
0   300 mls

MISSING

After piloting *Norge* across the Arctic, Nobile felt he had not received appropriate credit. So in May 1928, he flew to the Pole with a new airship, *Italia*. Tragically, on the return south *Italia* crashed, leaving nine men stranded on the ice. An international rescue effort ensued, in the process of which Amundsen vanished into the unknown. Finally, in late June a Swedish pilot discovered the party and brought back the injured Nobile. The others were rescued off the ice in July by the Soviet icebreaker *Krasin*.

LEFT The dirigible *Italia*, which crashed on its return from the North Pole. Nobile was severely criticized for returning on the rescue plane before his men.

UNDER THE POLE

At the height of the Cold War, with the US and USSR nervously watching each other across the Arctic Basin, the Americans made a technological statement. On 3 August 1958, the nuclear-powered submarine *Nautilus*, under the command of John Anderson and with 116 men aboard, cruised under the North Pole at a depth of 120 metres (400 feet). Another US submarine, *Skate*, commanded by James Calvert, soon followed, surfacing some 65 kilometres (40 miles) from the Pole on 12 August. On 17 March 1959, Skate returned to become the first submarine to surface at the Pole.

ABOVE The key figures in the first attainment of the North Pole. From left: Umberto Nobile (with his dog Titina), the designer and pilot of *Norge*; Lincoln Ellsworth, the American aviator who provided key funding; Roald Amundsen, commander and public face of the expedition; and Hjalmar Riiser-Larsen, second-in-command and navigator. There were power-struggles over the crew's composition, as Amundsen and Nobile wanted their respective countries to have the most men.

ABOVE RIGHT Richard E. Byrd became an American hero after claiming to have reached the North Pole in 1926. Although his claims were later shown to be false, his fame allowed him to make impressive accomplishments on several Antarctic expeditions.

RIGHT Modern technology meets exploration. This record was released as a highlight of the cruise of USS *Nautilus* under the North Pole.

tri-motor Fokker and even gave him assistance. On 9 May 1926, Byrd and his pilot Floyd Bennett took off, returning less than 16 hours later, claiming to have reached the Pole. Amundsen, Ellsworth and Nobile then flew to the Pole on 12 May, and continued to Teller, Alaska, completing the first crossing of the Arctic Basin. Amundsen had clearly passed over the North Pole, but serious doubts later arose as to whether Byrd had really had time to reach it. Then in 1996, Byrd's original record of the trip was located, which indicated that he *had not* reached the Pole, and knew that he had not. So when all was done, Amundsen had been the first to see the North Pole.

But who was the first actually to stand there? That honour belongs to a group of 24 Russians, led by Aleksandr Kuznetsov, who on 23 April 1948, landed there in three aeroplanes. The hydrologist Pavel Gordiyenko has been mentioned as the first to disembark, but that remains uncertain. It was to be another 20 years – almost six decades after Cook and Peary – before the North Pole was actually reached via the surface. On 19 April 1968, on his second attempt, Minnesota insurance agent Ralph Plaisted arrived at the Pole on a snowmobile, having left from Ward Hunt Island. Almost a year later, on 5 April 1969, while making the first surface crossing of the Arctic Ocean, Wally Herbert and several companions became the first men to attain the Pole the traditional way: by dog sledge.

ANTARCTICA

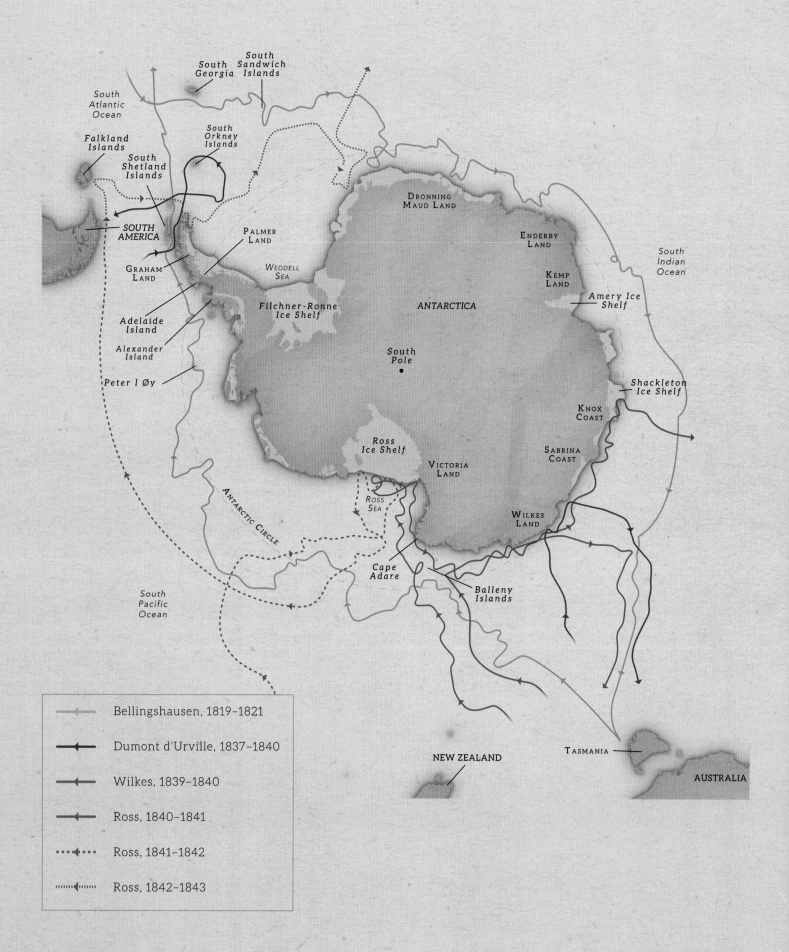

South
Georgia

South
Sandwich
Islands

South
Atlantic
Ocean

South
Orkney
Islands

Falkland
Islands

South
Shetland
Islands

SOUTH
AMERICA

PALMER
LAND

GRAHAM
LAND

Weddell
Sea

Adelaide
Island

Alexander
Island

Peter I Øy

ANTARCTIC CIRCLE

South
Pacific
Ocean

Filchner-Ronne
Ice Shelf

DRONNING
MAUD LAND

ENDERBY
LAND

KEMP
LAND

Amery Ice
Shelf

South
Indian
Ocean

ANTARCTICA

South Pole

South
Pole

Shackleton
Ice Shelf

KNOX
COAST

SABRINA
COAST

Ross
Ice Shelf

VICTORIA
LAND

Ross
Sea

WILKES
LAND

Cape
Adare

Balleny
Islands

NEW ZEALAND

TASMANIA

AUSTRALIA

Bellingshausen, 1819–1821

Dumont d'Urville, 1837–1840

Wilkes, 1839–1840

Ross, 1840–1841

Ross, 1841–1842

Ross, 1842–1843

THE GREAT NATIONAL EXPEDITIONS

Four decades after Captain Cook made the first circumnavigation of Antarctica, a Russian naval expedition under the command of Estonian-born Fabian von Bellingshausen repeated the feat. The exploration of the Antarctic had recently been dominated by sealers, but in 1819 Bellingshausen sailed from Kronstadt in command of the ships *Vostok* and *Myrnyy*. His official objectives included approaching the South Pole, improving the maps of the southern oceans, and carrying out a wide range of scientific studies.

TOP Dumont d'Urville sailed twice around the world before being placed in charge of the French Antarctic expedition. His ship, Astrolabe, was caught in the ice In the Weddell Sea before being freed.

ABOVE Wilkes was a junior lieutenant when he was put in command of the United States Exploring Expedition. His achievements were considerable, but he was an unpopular leader.

RIGHT Adélie penguins were first recorded by Dumont d'Urville, who named them after his wife. They are one of three species of pygoscelid (brush-tailed) penguins; the others are gentoo penguins and chinstrap penguins.

In early January 1820, the Russians discovered the Traversay Islands, the northern group of the South Sandwich Islands. In the next three months they made six attempts to reach the South Pole but were halted each time by the ice, although they made the first ever sighting of an ice shelf and then the Antarctic mainland itself. The expedition continued around the continent, in the process discovering Peter I Øy and Alexander Island.

Sealers again dominated the exploration of the Antarctic until the late 1830s, when three national expeditions were launched. Each was driven in part by a desire to attain the South Magnetic Pole, the location of which had been predicted by the German geographer Karl Friedrich Gauss. In 1837–40 Jules-Sébastien-César Dumont d'Urville, a cultured, outspoken naval officer who had earlier persuaded the French government to purchase the *Venus de Milo*, sailed two corvettes, *Astrolabe* and *Zelée*, first into the Weddell Sea and then, after a year and a half in the Pacific, into the little-known regions south of Australia. Here D'Urville discovered the area he named Terre Adélie after his wife – as he did Adélie penguins – following which, having found his way towards the Magnetic Pole blocked by ice, he headed north.

On 29 January 1840, in remote Antarctic waters, one of the most remarkable encounters in the history of exploration took place. D'Urville spied a brig, which first headed rapidly towards them, then, misinterpreting a French action, veered away. It was USS *Porpoise*, one of six vessels originally making up the United States Exploring Expedition (1838–42), under the overall command of Lieutenant Charles Wilkes. Like the French, the Americans concentrated on the Pacific and Atlantic, but twice made Antarctic excursions. The first of these was into the Weddell Sea, the second skirting hundreds of miles along the coast of what became known as Wilkes Land. Although Wilkes produced the first charts to call the new land the "Antarctic Continent", he was widely criticized after later explorers sailed through areas he marked as land. In addition, his poor treatment of his officers and scientists led to his court-martial.

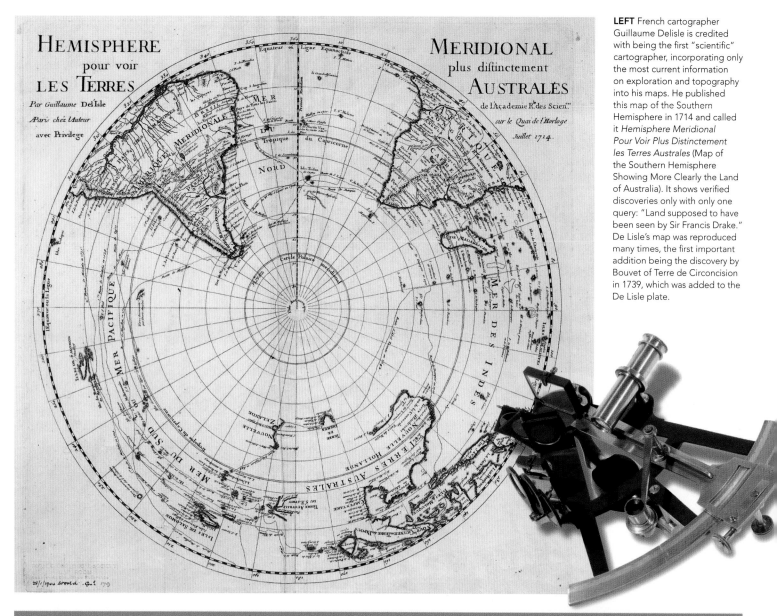

LEFT French cartographer Guillaume Delisle is credited with being the first "scientific" cartographer, incorporating only the most current information on exploration and topography into his maps. He published this map of the Southern Hemisphere in 1714 and called it *Hemisphere Meridional Pour Voir Plus Distinctement les Terres Australes* (Map of the Southern Hemisphere Showing More Clearly the Land of Australia). It shows verified discoveries only with only one query: "Land supposed to have been seen by Sir Francis Drake." De Lisle's map was reproduced many times, the first important addition being the discovery by Bouvet of Terre de Circoncision in 1739, which was added to the De Lisle plate.

THE GREAT ICE BARRIER

The Great Ice Barrier, today known as the Ross Ice Shelf, is one of the world's greatest natural wonders: a roughly triangular plate of floating ice covering about 527,000 km2 (203,000 square miles), making it larger than Spain or California. The average depth of the ice is approximately 370 metres (1,215 feet). "We might with equal chance of success," wrote Ross, "try to sail through the cliffs of Dover." In 1908–09 Ernest Shackleton's party first crossed the Barrier, and Roald Amundsen's expedition was later the first to winter on it.

LEFT The explorers who first saw the Ross Ice Shelf expected it to have a flat, even surface. How wrong they were! As seen here, the surface at varying points is broken, ridged, deeply crevassed, and incredibly dangerous.

The period's other major expedition (1839–43) was led by the Royal Navy's James Clark Ross, who, in 1831, had been the first to attain the *North* Magnetic Pole. Sailing from Van Diemen's Land (Tasmania) in HMS *Erebus* and HMS *Terror*, Ross proved that the southern ice pack could indeed be navigated, and that beyond it was an open sea, which allowed him to reach a farthest south of 78° 10' S. He discovered Victoria Land, several islands, including Ross Island (with the volcanoes Erebus and Terror), and the Great Ice Barrier – now known as the Ross Ice Shelf which he traced for some 600 kilometres (375 miles). The next year he again ventured into the Ross Sea before proceeding to the Falkland Islands and thence pushing into the Weddell Sea. Not only did Ross make great strides in magnetic research, but his assistant surgeon, Joseph Dalton Hooker, produced ground-breaking work on Antarctic botany.

Three decades after these efforts, another British expedition aboard HMS *Challenger* carried out a programme of exploration of the depths of the sea. Although the Antarctic continent was never sighted in the expedition's four years (1872–76), *Challenger* became the first steam vessel to cross the Antarctic Circle, new scientific knowledge was added about the Antarctic and the sub-Antarctic islands, and the science of oceanography was effectively born. The expedition also contributed to public interest in the far south, which helped set the scene for the "Heroic Age" of Antarctic exploration.

OPPOSITE CENTRE RIGHT
The sextant was widely used after Thomas Godfrey and John Hadley independently discovered its principles in 1730. Named for its original shape as a sixth of a circle, the sextant was used to determine latitude by measuring the angle subtended by a celestial body to the horizon.

RIGHT A letter from Jules-Sébastien-César Dumont d'Urville to Sir John Franklin, then lieutenant-governor of Van Diemen's Land, written while the expedition headed back to France. Dumont d'Urville's men had suffered dreadfully from a massive outbreak of dysentery in 1839, but had been treated most hospitably by Franklin and the authorities in Hobart, where they were nursed back to health. (See Translations, page 205.)

BELOW In four years HMS *Challenger* logged 110,900 kilometres (68,890 miles), while the scientists aboard made a systematic series of measurements at 362 observation stations. Many of the results were worked up in the onboard zoological laboratory.

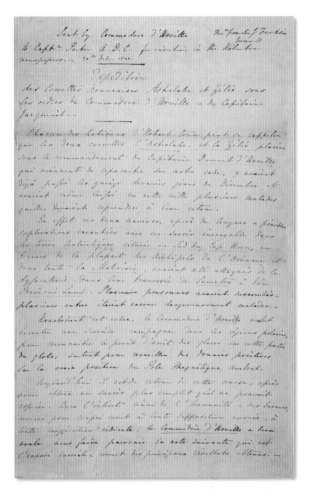

WHO FIRST SAW ANTARCTICA?

For years, Edward Bransfield was considered the first man to see the Antarctic continent, while exploring the South Shetland Islands. However, on 27 January 1820, Bellingshausen's expedition sighted the mainland, beating Bransfield by three days. But it took more than a century for this to be accepted – when Bellingshausen's account was translated into English. Nathaniel Palmer, an American sealer, was also supposed to have seen the Antarctic Peninsula from the South Shetlands in November 1820. But even if he did, it was more than nine months after Bellingshausen and Bransfield.

TOP Hut Point at the south end of McMurdo Sound, where the British National Antarctic Expedition set up its hut. The expedition ship *Discovery* can be seen behind the hut.

BOTTOM Ernest Shackleton, Robert Falcon Scott, and Edward Wilson before their departure onto the Great Ice Barrier. They established a farthest south but struggled mightily on their return due to scurvy and exhaustion. Scott then sent the invalid Shackleton home while other expedition members remained in the Antarctic.

THE FIRST WINTERERS

After the great national expeditions of the nineteenth century, there was little active exploration of the Antarctic for the next 50 years. But in the 1890s, whaling expeditions from Scotland and Norway went south to determine whether there were enough baleen whales to start an Antarctic industry. Accompanying these voyages were scientists who proved there was great value to Antarctic research. Then, in 1895, led by Clements Markham of the RGS, the Sixth International Geographical Congress passed a resolution stating that the Antarctic was the greatest remaining focus for geographical investigation. Science, commerce, and geographical exploration were a powerful trio, and together they launched the "Heroic Age of Antarctic Exploration".

TOP Carsten Borchgrevink was a Norwegian who had resettled in Australia. Making good use of his claim to have been the first man to set foot on the Antarctic continent, he was able to convince the publishing magnate Sir George Newnes to sponsor his expedition.

MIDDLE Erich von Drygalski, leader of a German Antarctic expedition (1901–03) which was caught in the ice for a year.

BOTTOM William S. Bruce, leader of the Scottish National Antarctic Expedition (1902–04). Because the expedition wasn't official Bruce never received the credit due to him in his lifetime

The first of these expeditions sailed in 1897 under Adrien de Gerlache on a former whaler renamed *Belgica*. They discovered new islands and passages west of the Antarctic Peninsula before being caught in the ice and drifting helplessly for a year, the men becoming the first to winter in the Antarctic. Confronted with massive uncertainties, they were held together by the mate, Roald Amundsen, who would later navigate the Northwest Passage and be the first man to reach the South Pole, and the doctor, Frederick Cook, who would become embroiled in a controversy over the attainment of the *North* Pole. *Belgica* was finally freed in March 1899 after escape was initiated by cutting a path through the ice to open water.

Later that year, another party became the first to winter on the continent proper. Led by Carsten Borchgrevink, who several years earlier had (inaccurately) claimed to be the first man to stand on the continent, the expedition returned to the same location where he had first landed: Cape Adare in Victoria Land. After conducting scientific investigations throughout the winter, Borchgrevink and two others skied south on the Great Ice Barrier, reaching a farthest south of 78° 50′ S and opening a path towards the Pole.

The next step in that direction came on the British National Antarctic Expedition (1901–04), led by Robert Falcon Scott. The specially built research ship *Discovery* wintered at the base of McMurdo Sound, from where the first extensive exploration of the continent was conducted. One party under Albert Armitage became the first to reach the Antarctic Plateau, while Scott, third officer Ernest Shackleton and surgeon Edward Wilson sledged deep onto the Great Ice Barrier, reaching a record latitude of 82° 16′ 33″ S. *Discovery* was trapped in the ice, so Scott remained another year before a relief expedition helped break them free in 1904. At the same time, other expeditions were dotted around the Antarctic. In 1902, a German effort under Erich von Drygalski became the second to winter in the pack ice, when its ship *Gauss* was trapped off an unknown Antarctic coastline. Held for an entire year, Drygalski made geographic observations from a balloon, and the scientists compiled vast quantities of new data, eventually filling 20 volumes of reports. The new region was named Wilhelm II Land, its outstanding feature being Gaussberg, an extinct volcano standing high above the icy coast. Meanwhile, both a Swedish expedition under Otto Nordenskjöld and the Scottish National Antarctic Expedition (1902–04) investigated the Weddell Sea region. From the ship *Scotia*, William

Speirs Bruce and his Scottish party discovered parts of Coats Land, but no landing was possible, so they wintered on Laurie Island in the South Orkneys. There, in 1903 Bruce established Omond House, a meteorological observatory that he subsequently turned over to Argentina; today it is the oldest continuously functioning station in Antarctica. Bruce's expedition recorded a wealth of geographical, meteorological and botanical information, and was probably the most profitable in oceanographic research of any early Antarctic expedition.

On the far side of the Antarctic Peninsula another remarkable effort was taking place. Having built the ship *Français* with his inheritance, Jean-Baptiste Charcot led a French expedition (1903–05) that discovered and charted new islands and coastline, while undertaking a comprehensive scientific programme. In 1908–10, Charcot returned to the same region in *Pourquoi Pas?* and achieved equally impressive results.

A TRAGEDY NARROWLY AVERTED

Perhaps the most dramatic early "Heroic Age" events came on Otto Nordenskjöld's (above) Swedish expedition. In 1902, his party was left at Snow Hill Island in the Weddell Sea, with Carl Larsen due to bring the ship *Antarctic* back in the spring. However, Larsen could not reach them, so he dropped three men at Hope Bay at the tip of the Peninsula to try. They could not reach Nordenskjöld either, and, meanwhile, *Antarctic* was crushed and sank; its crew walked across the ice to Paulet Island. After an incredibly hard winter, the Hope Bay party reached Nordenskjöld. In November 1903, an Argentine ship rescued the men at Snow Hill Island, and, by a stroke of good fortune, only hours before they departed, Larsen arrived, having walked over the ice to Hope Bay and thence to Snow Hill. The ship collected the crew at Paulet and steamed north to safety.

ABOVE Shackleton's hut at Cape Royds. Unable to reach Scott's base at Hut Point, Shackleton established his headquarters 18 miles up the coast of Ross Island, under the shadow of Mount Erebus.

BELOW LEFT On 4 February 1902, Shackleton took this picture of the balloon *Eva* on the Great Ice Barrier. Scott was the first man to go aloft in the Antarctic, and shortly thereafter Shackleton became the second. When asked if he wished to make an ascent, Wilson proclaimed it dangerously foolhardy.

BELOW RIGHT The scientific staff of Jean-Baptiste Charcot's first Antarctic expedition, which sailed in his own ship *Français*. Charcot, who was in charge of bacteriology, is in the centre of the front row.

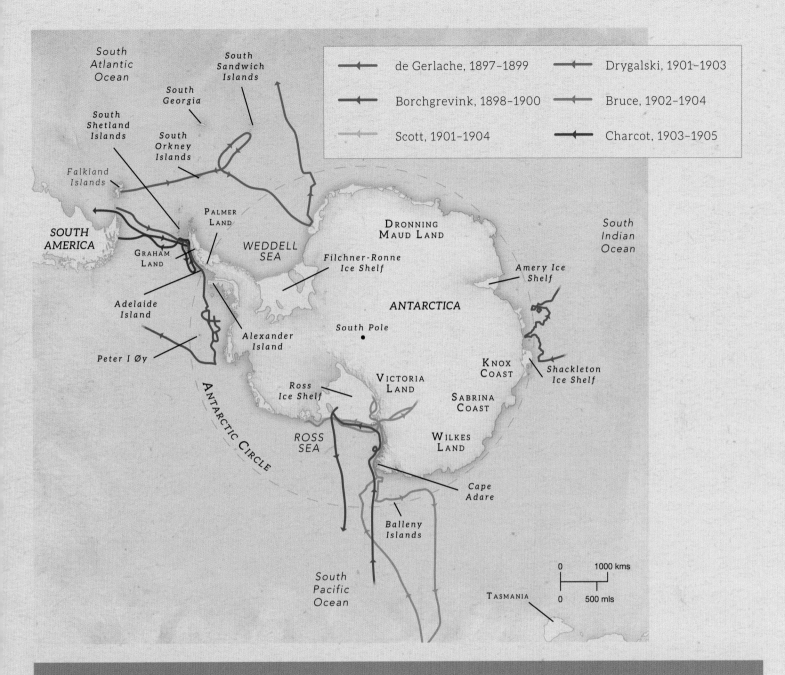

Legend:
- de Gerlache, 1897–1899
- Borchgrevink, 1898–1900
- Scott, 1901–1904
- Drygalski, 1901–1903
- Bruce, 1902–1904
- Charcot, 1903–1905

South Atlantic Ocean

South Sandwich Islands

South Georgia

South Shetland Islands

South Orkney Islands

Falkland Islands

SOUTH AMERICA

PALMER LAND

GRAHAM LAND

Adelaide Island

Peter I Øy

Alexander Island

WEDDELL SEA

DRONNING MAUD LAND

Filchner-Ronne Ice Shelf

ANTARCTICA

South Pole

ANTARCTIC CIRCLE

Ross Ice Shelf

VICTORIA LAND

ROSS SEA

Cape Adare

Balleny Islands

South Pacific Ocean

South Indian Ocean

Amery Ice Shelf

KNOX COAST

Shackleton Ice Shelf

SABRINA COAST

WILKES LAND

TASMANIA

0 — 1000 kms
0 — 500 mls

SEALERS, WHALERS, AND ANTARCTIC DISCOVERY

Much early Antarctic exploration was carried out by sealers and whalers looking to satisfy the market for oil, fur and baleen. Although some exploration went unreported – so that other hunters could not poach the resources – major discoveries included Macquarie Island in 1810 (Frederick Hasselburg), the South Orkney Islands in 1821 (George Powell and Nathaniel Palmer), the Weddell Sea in 1823 (James Weddell) and the Balleny Islands in 1839 (John Balleny). The first confirmed landing on the continent was also made by a sealer, John Davis from New Haven, Connecticut, who landed at Hughes Bay, Antarctic Peninsula, in 1821.

LEFT James Weddell, a sealer with strong scientific interests, sailed in the brig *Jane*, to a farthest south of 74°15′ S in what is now named the Weddell Sea.

Reference to Contours
Showing depths in fathoms

Sea level

100 Fathoms

500 "

1000 "

2000 "

3000 "

Below 3000

Present Antarctic Continent,

With South Polar Regions drawn to illustrate the probable topography
as well as the effects of Messrs Amundsen and Shirase's
Expeditions to the Antarctic.

By

M. Ikeda "Nasakinski"

(Chief Scientist to M. Shirase's Antarctic Expedition)

PACIFIC

OCE

Probably
dense
pack ice

probably barrier
with islands or
undulating land

King Edward VII
Land

Mt. Kobu.
Mt. Ōkuma
Queen Maud Range

SOUTH
Pole
POLE

Chatham Is

Bounty I

Antipodes I

C Adare

Ross Sea

Erebus

15100

Wellington Strait

Port Lyttelton

Campbell I

Balleny
Is

South

AUCKLAND

NEW
ZEALAND

Auckland
I

Macquaries I

Victoria
Land

Adelie
Land

Clarie
Land

Lord Howe I

Sabrina
Land

Knox
Land

Sydney

Tasmania

Bass Strait

Melbourne

AUSTRALIA

Adelaide

Gt Australian
Bight

Albany

INDIA

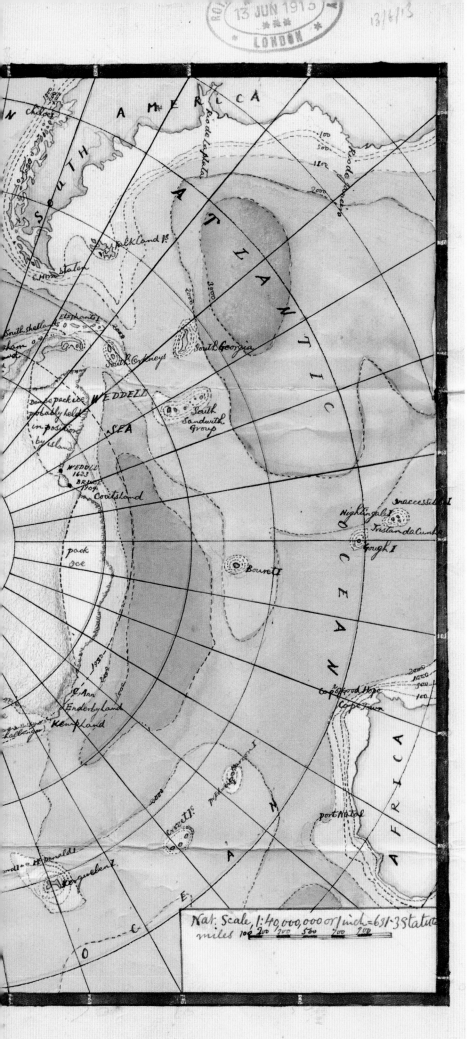

LEFT Map of Antarctica showing routes of Amudsen's and Shirase's expeditions, 1912.

ASSAULT ON THE POLE

Following the farthest south reached by Robert Falcon Scott, Ernest Shackleton and Edward Wilson on the *Discovery* Expedition, Shackleton's debilitated condition led to Scott invaliding home the junior officer in March 1903, a year before the expedition returned. For several years thereafter Shackleton hoped to make an assault on the South Pole, not only to overcome any accusations of personal weakness, but also to accomplish something that would lead to fame and fortune. In 1907, a £7,000 loan from the Scottish industrialist William Beardmore allowed him to proceed. Early the next year, Shackleton's party of 15 established a base at Cape Royds on Ross Island; while there, they made the first ascent of Mount Erebus, the island's active volcano.

The next spring, Shackleton's men made two major treks. T. W. Edgeworth David, Douglas Mawson and Alistair Mackay man-hauled approximately 2,000 kilometres (1,250 miles) to the vicinity of the South Magnetic Pole. Meanwhile, Shackleton, Frank Wild, Eric Marshall and Jameson Adams headed for the South Pole, first with Manchurian ponies and then by man-hauling. They crossed the Great Ice Barrier, ascended the Beardmore Glacier to the Antarctic Plateau, and continued on until forced to retreat by lack of food. Nevertheless, at 88° 23′ S they were only 180 km (97 nautical miles) short of the South Pole and had made one of the epic journeys in polar history.

Soon after Shackleton's return, Scott announced an expedition with dual aims: conducting a broad scientific programme and attaining the Pole. In early 1911, Scott's men established two bases, at Cape Adare and at Cape Evans on Ross Island. Late in the process, Scott found that he was not alone in his quest. Roald Amundsen, the conqueror of the Northwest Passage, had initially hoped to reach the *North* Pole. But when Frederick Cook and Robert E. Peary claimed it, he

changed his goal to the *South* Pole and camped on the Great Ice Barrier.

Both men were desperate to be first, and, after the winter and an abortive start, Amundsen's five-man, 52-dog party headed south on 20 October 1911. They soon went through virgin territory, then ascended to the Plateau via the treacherous Axel Heiberg Glacier, near the top of which 24 dogs were killed to provide food for the men and other dogs. On 14 December Amundsen, Helmer Hanssen, Olav Bjaaland, Oscar Wisting and Sverre Hassel reached the Pole, and several days later they began their return to the awaiting *Fram*.

Scott, meanwhile, planned his transport around not only dogs, but ponies, motor sledges and man-hauling. After a winter that included ongoing

TOP Scott, Bowers, Wilson and Evans stand dejectedly at the Norwegian tent as the full impact of Amundsen's priority sinks in.

ABOVE An ice thermometer used on Scott's British National Antarctic Expedition of 1901–04. Second officer Michael Barne was in charge of taking sea temperatures, depth soundings and tidal measurements.

ABOVE & RIGHT The two competitors in the "Race for the Pole": Roald Amundsen and Robert Falcon Scott. Amundsen – 39 when he became the first to reach the South Pole – was the consummate planner, and his years in the Arctic during the navigation of the Northwest Passage helped show him the advantages of dressing in Eskimo clothing and using dogs and skis for transportation.

The 43-year-old Scott was an intriguing combination of conventional and modern, his transportation including dogs, ponies, new designs in motor transport, and traditional man-hauling.

BELOW Shirase's "Dash Patrol" was the first party to explore King Edward VII Land, which earlier explorers had been unable to reach.

SHIRASE'S JAPANESE ANTARCTIC EXPEDITION

In December 1910, after lengthy efforts to gain support for an Antarctic expedition, Nobu Shirase sailed south from Japan. Heavy ice in the Ross Sea prevented the tiny *Kainan Maru* from reaching the Antarctic coast, so Shirase's party headed for Sydney, where their shoestring budget forced them to spend the winter almost as paupers. Back to the Great Ice Barrier in early 1912, Shirase led the "Dash Patrol" 250 kilometres (160 miles) south. Rough conditions made boarding the ship hazardous on their return, but, that accomplished, they reached home as heroes.

scientific research and a nightmare trek by Wilson, "Birdie" Bowers and Apsley Cherry-Garrard to an emperor penguin rookery, Scott led a large party south. His plan incorporated assistance from numerous support parties, which he sent back at different stages, the final one returning from the Plateau, while Scott, Wilson, Bowers, Edgar Evans and L. E. G. Oates followed Shackleton's path.

The five men reached the South Pole on 17 January 1912 to find that Amundsen had beaten them. Their return was plagued by disappointment, inadequate food and fuel, freakishly cold weather and incipient scurvy. Evans died near the base of the Beardmore Glacier in mid-February. A month later, Oates perished in an heroic effort to save his comrades, following which Scott, Wilson and Bowers reached a point 18 kilometres (11 miles) from the well-stocked One Ton Depot before the weather forced them to remain in their tent, where all three died. The next spring, other expedition members found their bodies.

In 1928, Richard E. Byrd, who had become famous when he claimed to have flown to the North Pole, led a large expedition that established a base named "Little America" near Amundsen's old base. After an extensive aerial reconnaissance, on 28–29 November 1929, Byrd and three others made the first flight to the South Pole, *en route* jettisoning much of their food in order to gain the altitude to pass over the Transantarctic Mountains. In just 19 hours they made a journey that had taken Amundsen three months and had doomed Scott's party.

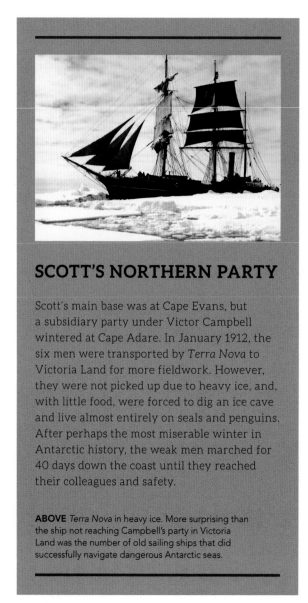

SCOTT'S NORTHERN PARTY

Scott's main base was at Cape Evans, but a subsidiary party under Victor Campbell wintered at Cape Adare. In January 1912, the six men were transported by *Terra Nova* to Victoria Land for more fieldwork. However, they were not picked up due to heavy ice, and, with little food, were forced to dig an ice cave and live almost entirely on seals and penguins. After perhaps the most miserable winter in Antarctic history, the weak men marched for 40 days down the coast until they reached their colleagues and safety.

ABOVE *Terra Nova* in heavy ice. More surprising than the ship not reaching Campbell's party in Victoria Land was the number of old sailing ships that did successfully navigate dangerous Antarctic seas.

BELOW A photo taken on the Polar Plateau by "Birdie" Bowers of his four sledging comrades on their return from the South Pole. From left: Evans, Oates, Wilson and Scott.

ABOVE Amundsen's party takes its formal leave of the tent placed where their calculations showed the South Pole to be, and that they named "Polheim". From left: Amundsen, Hanssen, Hassel and Wisting.

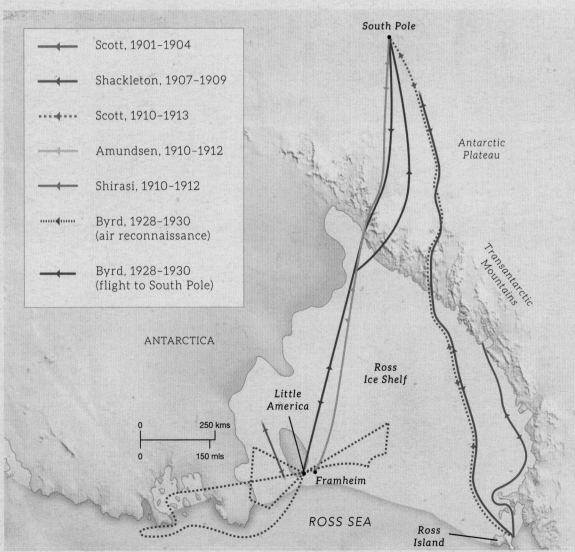

Scott, 1901–1904

Shackleton, 1907–1909

Scott, 1910–1913

Amundsen, 1910–1912

Shirasi, 1910–1912

Byrd, 1928–1930
(air reconnaissance)

Byrd, 1928–1930
(flight to South Pole)

South Pole

Antarctic Plateau

Transantarctic Mountains

ANTARCTICA

Ross Ice Shelf

Little America

0   250 kms

0   150 mls

Framheim

ROSS SEA

Ross Island

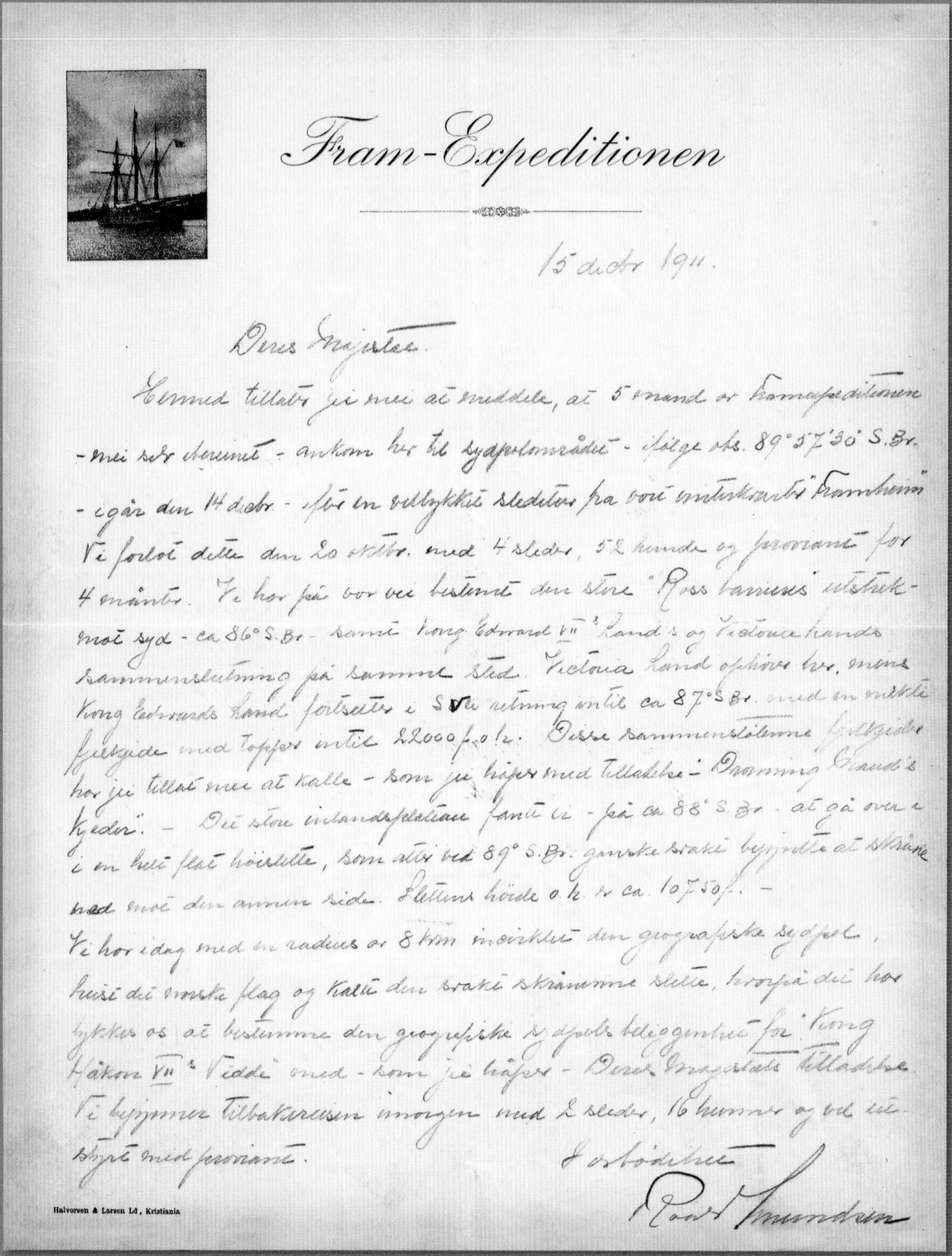

Fram-Expeditionen

15 desbr 1911.

Deres Majestæt.

Hermed tillater jei mei at meddele, at 5 mand av Framexpeditionen – mei selv iberegnet – ankom her til sydpolområdet – ifølge obs. 89°57'30" S.Br. – igår den 14 desbr. – efter en vellykket sledetur fra vort vinterkvarter "Framheim". Vi forlot dette den 20 oktbr. med 4 sleder, 52 hunde og proviant for 4 måneder. Vi har på vor vei bestemt den store "Ross barrierens" utstrækning mot syd – ca 86° S.Br. – samt Kong Edward VII's Land's og Victoria Lands sammenslutning på samme sted. Victoria Land ophører her, mens Kong Edwards Land fortsætter i SØ retning intil ca 87° S.Br. med en mektig fjeldside med topper intil 22000 f.o.h. Disse sammenslutninger fjeldsider har jei tillat mei at kalde – som jei håper med tilladelse – "Dronning Maud's Kjeder". – Det store indlandsplateau fandt vi – på ca 88° S.Br. – at gå over i en helt flat høislette, som atter ved 89° S.Br. ganske svakt begyndte at skråne ned mot den anden side. Slettens høide o.h. er ca. 10750 f. –

Vi har idag med en radius av 8 km indsirklet den geografiske sydpol, heist det norske flag og kaldt den svakt skrånende slette, hvorpå det har lykkes os at bestemme den geografiske sydpols beliggenhet for "Kong Håkon VII's Vidde" med – som jei håper – Deres Majestæts tilladelse.

Vi begynner tilbakereisen imorgen med 2 sleder, 16 hunde og vel utstyrt med proviant.

Ærbødigst

Roald Amundsen

Halvorsen & Larsen Ld, Kristiania

LEFT A letter written by Roald Amundsen to King Haakon VII of Norway the day after his party became the first ever to reach the South Pole. Amundsen left the letter in the tent they called "Polheim" with a request that Scott take it back in case Amundsen's party failed to return safely. It was found with the bodies of Scott and his companions, proving beyond doubt that the Norwegians had beaten Scott's party to the Pole. (See Translations, page 205)

TELEPHONE.
No. 67 MAYFAIR.

ROYAL GEOGRAPHICAL SOCIETY
1830

1, Savile Row,
Burlington Gardens,
London, W.

March 26th, 1912.

Capt. Roald Amundsen,
    C/o. Norwegian Consul,
       Buenos Aires, Argentine Republic.

Dear Capt. Amundsen,

        Our President, Lord Curzon has cabled to you the
hearty congratulations of the Society on the magnificent
journey you made to the South Pole and back.   I send you my
own personal warm congratulations, although I hope you will
allow me to say that I wish you had been quite frank about your
intentions before you left.   Captain Scott would have been the
last man in the world to object to your making an attempt, to
attain the South Pole. *But it 's no use discussing this now*
~~I think you were mistaken.~~   However,
~~it is all over now and you have been successful.~~   I hope you
will reap the rich reward that you deserve.

        I learn from the lecture Agent here, Mr Christy,
that you propose to come to England next autumn unless Captain
Scott returns home in the meantime and intends to lecture in
England, in which case you would not interfere.   Should you
come to England our Council would be glad if you would give

ABOVE The first page of a letter from J. Scott Keltie, the Secretary of the Royal Geographical Society, to Roald Amundsen, inviting the explorer to give a lecture at the RGS. The tone of the letter leaves no doubt about the way Keltie – and most of the British geographical establishment – felt about Amundsen beating Robert Falcon Scott to the South Pole.

RIGHT A list of supplies purchased from Harrod's for *Morning*, the ship sent to relieve Robert Falcon Scott's British National Antarctic Expedition in 1903. It is remarkable to note what £2,326, 9 shillings, 3½ pence could buy.

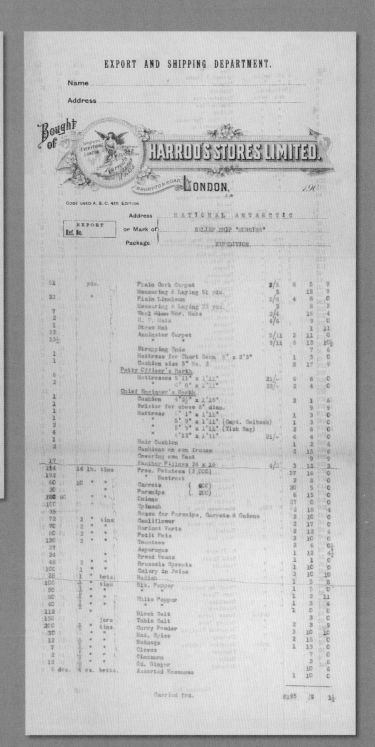

DOUGLAS MAWSON

Although arguably the greatest Antarctic scientist, Douglas Mawson came to the field quite by chance. Born in Yorkshire in 1882, he grew up in New South Wales, and at the University of Sydney became a protégé of the famed geologist T. W. Edgeworth David. In 1907, David was given the opportunity to sail to Antarctica with Ernest Shackleton's first expedition and then return after the ship left behind a wintering party. Mawson, then a lecturer at the University of Adelaide, asked Shackleton if he could do likewise, and was surprised to be named the physicist for the entirety of the expedition.

OPPOSITE TOP A snow tunnel dug to enable Mawson's men to leave the hut. The wind blew so much drift in that it piled up to the roof tops. On the right is Belgrave Ninnis.

OPPOSITE LEFT At the last minute, Mawson consolidated two parties into one at his Main Base at Cape Denison. When completed, the two huts stood adjacent to each other, one serving as living quarters and the other as a work area.

OPPOSITE RIGHT The remains of Mawson's sledge after he reached Cape Denison. When Xavier Mertz died, Mawson cut it in half with a pen knife so that he could haul it.

ABOVE A portrait of Douglas Mawson taken during the preparations for his AAE. The greatest scientific expedition in the history of the Antarctic was strictly his brainchild.

RIGHT In 1909, Alistair Mackay, Edgeworth David and Mawson claimed the region of the South Magnetic Pole for the British Empire. The string pulling the camera shutter can be seen extending from David's hand.

Mawson proceeded to play a key role. In the expedition's early stages, he was a member of the party that made the first ascent of Mount Erebus, the active volcano on Ross Island. The next spring, he, David – who also remained for the duration – and surgeon Alistair Mackay man-hauled approximately 2,000 kilometres (1,250 miles), becoming the first men to reach the vicinity of the South Magnetic Pole, although their scientific instruments did not allow them to define its precise location. Mawson recognized the huge value of the scientific knowledge gained, and by the time the expedition returned north he was already considering continuing his Antarctic research.

Mawson's goal was to investigate the 3,200-kilometre- (2,000-mile-) long coast of Antarctica directly south of Australia, then a virtually unknown area. To do this, his Australasian Antarctic Expedition (AAE) included the most extensive scientific plan that had ever been envisaged for the Antarctic. Difficulties with finding landing sites in early 1912 meant last-minute changes, but the AAE nevertheless comprised three stations: one on sub-Antarctic Macquarie Island (with five men), the Main Base at Cape Denison, Commonwealth Bay (18 men), and the Western Base on the Shackleton Ice Shelf (eight men). From these – and on several cruises aboard the expedition ship *Aurora*, commanded by John King Davis – the expedition members conducted the most extensive Antarctic scientific program that had yet been carried out. The two parties on the mainland also explored widely on sledging trips, and members from Western Base reached Gaussberg, discovered by Erich von Drygalski in 1902.

Difficulties were soon encountered at Main Base, because Mawson had unknowingly selected the windiest place on Earth for his station. For months

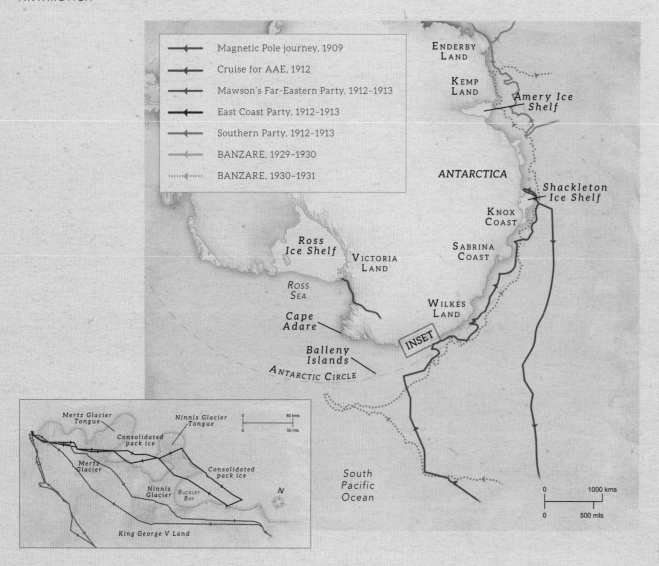

on end, the average wind speed for every hour of every day exceeded 80 km/h (50 mph). There were days when it averaged more than 145 km/h (90 mph), and it peaked at more than 320 km/h (200 mph). Nevertheless, the next spring Mawson sent six sledging parties in different directions, taking magnetic, meteorological and other measurements, charting coastlines and discovering new land. One party, which included photographer Frank Hurley, became the second to reach the region of the South Magnetic Pole, although lack of food meant they had to turn back short of the Pole itself.

The most notable sledging trip, however, was Mawson's. Some 500 kilometres (310 miles) from base, one of his two companions, B. E. S. Ninnis, was killed in a fall down a crevasse; the best dogs and most of the food were lost with him. Mawson and Xavier Mertz made a desperate dash across the icecap for Cape Denison, eating the remaining dogs

as they went. Both men suffered dreadfully from vitamin-A poisoning, caused by eating the dogs' livers, and the condition contributed to Mertz's death. Mawson continued alone and, after perhaps the most remarkable journey in Antarctic history, eventually struggled into base, only to find that *Aurora* had left hours before, and that he had to remain in the Antarctic for another year, along with six men who had been left behind to search for him.

Mawson returned to Australia a national hero and settled down to a scholarly career. His Antarctic interests continued, however, and in 1929–31 he led the British, Australian and New Zealand Antarctic Research Expedition (BANZARE), on which he discovered new coastal regions, charted others from ship and aircraft and made territorial claims that led to the establishment of the Australian Antarctic Territory. Mawson remained a key figure in Antarctic research until his death in 1958.

THE GREATEST ANTARCTIC SEA CAPTAIN

In 1907, on a whim, a young sailor named John King Davis applied as first mate on Shackleton's *Nimrod*. He got the job and performed so well that he was named captain for the voyage back to England. Davis then served as captain of *Aurora* and second-in-command of Mawson's AAE, during which he made three voyages to the Antarctic and two more to the sub-Antarctic. After commanding the relief expedition for Shackleton's Ross Sea party, Davis captained *Discovery* on Mawson's BANZARE, finishing his career as the most experienced of all Antarctic sea captains.

LEFT John King Davis was one of the most intense and demanding captains imaginable. He was so single-minded and so totally eschewed frivolity aboard that he was widely called "Gloomy Davis".

ANTARCTIC PHOTOGRAPHY

In the history of Antarctic photography, two names stand above all others. Englishman Herbert Ponting, of Robert Falcon Scott's last expedition, was a loner who rarely helped with other work. However, at his trade he was the supreme perfectionist, determining a photograph's best composition and then waiting endlessly for the ideal moment. Conversely, Australian Frank Hurley, who accompanied Mawson and then Shackleton, engaged in sledging and other expedition tasks, but still produced stunning photographs. Although an opportunist who photographed events as they occurred, Hurley had an uncanny ability to create incredibly evocative pictures.

ABOVE Frank Hurley's photo of his colleagues collecting ice to melt for water in the hut at Cape Denison. Virtually every time the men went outside they faced blizzard conditions.

ABOVE AND TOP LEFT Frank Hurley served under Mawson on both the AAE and BANZARE, but perhaps his most famous pictures came from his time on Shackleton's Imperial Trans-Antarctic Expedition, on which he photographed this unique ice formation.

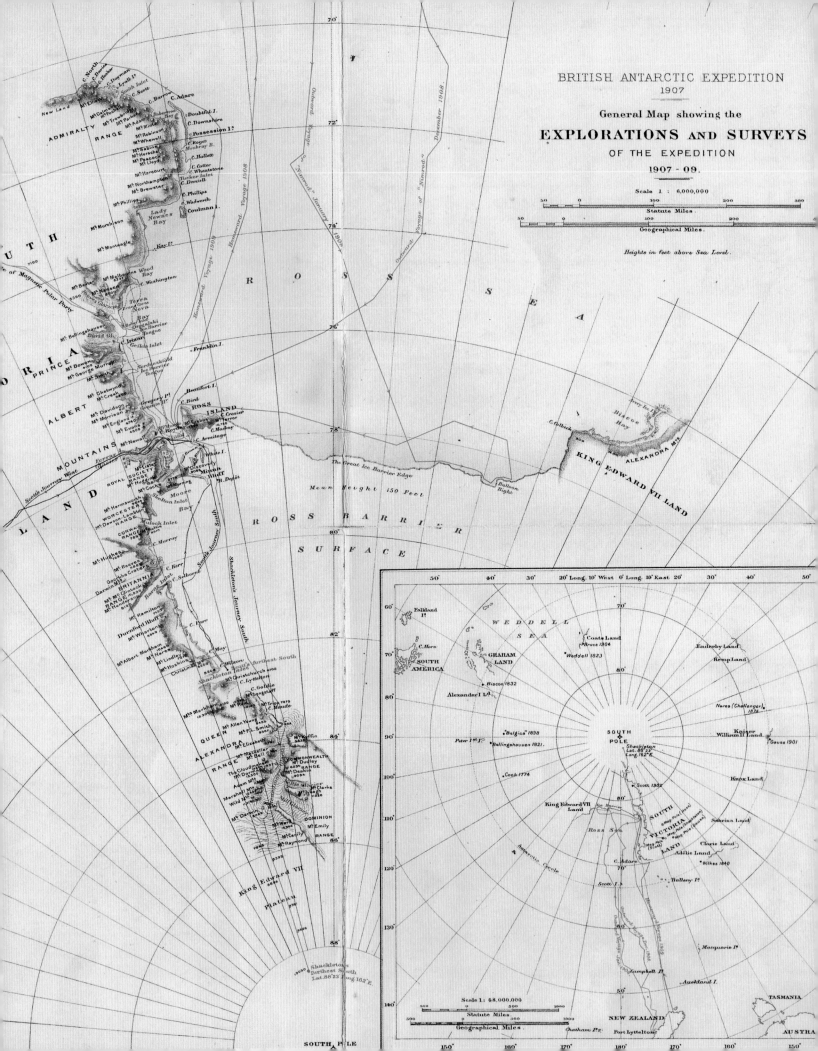

BRITISH ANTARCTIC EXPEDITION
1907

General Map showing the

EXPLORATIONS AND SURVEYS

OF THE EXPEDITION

1907-09.

Scale 1 : 6,000,000

Statute Miles.

Geographical Miles.

Heights in feet above Sea Level.

CROSSING ANTARCTICA

The South Pole had not even been reached before explorers began proposing a crossing of the Antarctic continent. In 1908, the Scottish oceanographer W. S. Bruce published such a plan, but lack of funding saw it still-born. Concurrently, the German explorer Wilhelm Filchner proposed an expedition in which two parties would meet in the continent's centre to determine if the Weddell and Ross seas were separated by a continuous landmass or an ice-covered series of islands. Lack of funding plagued Filchner, too, and his goal was reduced to a one-way crossing of the continent.

OPPOSITE Ernest Shackleton organized and led the *Nimrod* Expedition (British Antarctic Expedition 1907–9) as an attempt to reach the South Pole and also to continue the scientific and geological survey work of the *Discovery* Expedition (led by Scott in 1901–4). Two separate parties headed out from base camp; the first discovered the South Magnetic Pole in January 1909. Shackleton was in the second party that attempted to reach the Geographic South Pole. They came within 97 miles (180km) of their goal but, hampered by poor weather, they made the brave decision to turn back on 9 January 1909, rather than risk almost certainly running out of food and fuel on the return journey.

ABOVE Wilhelm Filchner experienced on the Filchner Ice Shelf what Shackleton had feared about establishing base on the Ross Ice Shelf – losing his camp on a calving glacier. Despite this, Filchner charted previously unknown coasts, conducted a mid-winter sledging journey and twice visited South Georgia.

In 1911, Filchner entered the Weddell Sea. He reached land near its base and discovered the massive ice shelf later named in his honour. In February 1912, he started to build a winter camp, but eight days later an enormous section of the ice shelf calved off, taking the camp with it. Although the men and most of the supplies were saved, the ship *Deutschland* was beset for nine months. When finally released, they had drifted more than 10° north, and Filchner abandoned further attempts.

However, his plan was not forgotten. Ernest Shackleton borrowed liberally from both Filchner and Bruce in developing his proposals for an Antarctic crossing. Shackleton's notion was to lead a party from the Weddell Sea across unknown regions to the South Pole, from where he would follow his own earlier route to the Ross Sea. Meanwhile, depots would be laid from Ross Island to the Beardmore Glacier to sustain his crossing.

When Shackleton sailed south in *Endurance* in late 1914, he had even worse luck than Filchner. The ship was caught in the ice of the Weddell Sea and after drifting helplessly for nine months was crushed. The expedition members camped on the ice for another six months before taking to three open boats, in which they reached Elephant Island. Frank Wild

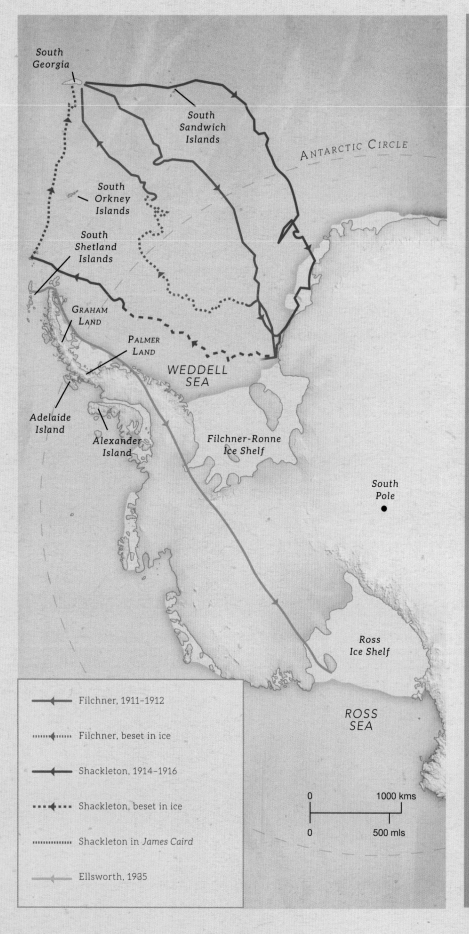

South
Georgia

South
Sandwich
Islands

ANTARCTIC CIRCLE

South
Orkney
Islands

South
Shetland
Islands

GRAHAM
LAND

PALMER
LAND

WEDDELL
SEA

Adelaide
Island

Alexander
Island

Filchner-Ronne
Ice Shelf

South
Pole

Ross
Ice Shelf

ROSS
SEA

- ◄—— Filchner, 1911–1912

- •••••◄•••• Filchner, beset in ice

- ◄—— Shackleton, 1914–1916

- ••◄••• Shackleton, beset in ice

- ••••••——— Shackleton in *James Caird*

- ◄—— Ellsworth, 1935

0   1000 kms

0   500 mls

MODERN ANTARCTIC ADVENTURERS

In recent decades, men and women have performed remarkable feats in Antarctica. Perhaps the best known is Ranulph Fiennes, whose numerous journeys include the Trans-Globe Expedition (1979–83) that circled the Earth through both Poles. In 1989–90, a six-man team including American Will Steger and Britain's Geoff Somers made the longest Antarctic crossing, 6,020 kilometres (3,741 miles) in 220 days with dog teams. Peter Hillary has lived up to his father's reputation with a series of Antarctic and mountain adventures. And Ann Bancroft and Liv Arneson were the first women to cross Antarctica in 2000–01.

ABOVE Sno-cat tracked vehicle in a typically precarious position over a deep crevasse during the crossing of Antarctica on the Commonwealth Trans-Antarctic Expedition.

ABOVE When Vivian Fuchs (centre) reached the South Pole while crossing Antarctica, he was greeted by Sir Edmund Hillary (left) and Rear Admiral George Dufek, commander of the American "Operation Deep Freeze".

OPPOSITE TOP One of the greatest living adventurers is Sir Ranulph Fiennes, whose daring accomplishments have taken him all over the world. His most impressive triumphs, however, have taken place in the Antarctic.

was left in charge, while Shackleton and five others sailed the whaleboat *James Caird* 1,450 kilometres (900 miles) to South Georgia. There, Shackleton, Frank Worsley and Tom Crean made a remarkable crossing of the island's unexplored mountain range to the Stromness whaling station. In the following months, Shackleton organized four successive relief expeditions and finally rescued his men, after they had been stranded on Elephant Island for 105 days.

Meanwhile, Shackleton's depot-laying party experienced equally difficult times. In May 1915, the expedition ship *Aurora* was driven from her moorings

in a blizzard, leaving 10 men behind without adequate supplies. Beset in the ice, she drifted for 10 months before limping to New Zealand. A subsequent relief expedition under John King Davis rescued the men left behind, reaching them in January 1917. In the interim, despite a lack of equipment, clothes and food, the party had laid supply depots for Shackleton all the way to the Beardmore Glacier. In the process, one man died of scurvy and two were lost on the sea ice.

Fifteen years later, Lincoln Ellsworth, who had financed Amundsen's Arctic flights, made three attempts to cross Antarctica by aeroplane. The first two were unsuccessful, but on the third, in November and December 1935, Ellsworth and Herbert Hollick-Kenyon made it the entire way – with a little help from their legs. What Ellsworth had figured as a 14-hour flight took 22 days, as refuelling and bad weather forced them to land four times. Then they ran out of fuel 26 kilometres (16 miles) short of Little America, to which they walked before being picked up by Ellsworth's ship.

The first land crossing of the continent was not made for three more decades, when Sno-cats, tractors and aerial reconnaissance finally made it possible. Beginning in 1955, the Commonwealth Trans-Antarctic Expedition, led by Vivian Fuchs, set up a station on the Filchner Ice Shelf, established a forward base, and then, in 1957–58, crossed the continent to the South Pole, where Fuchs's party was met by Edmund Hillary's New Zealand support team. Fuchs then successfully continued to Scott Base on Ross Island, near where Shackleton had hoped to finish so many years before.

FRANK WILD

No explorer of the "Heroic Age" spent more time in Antarctica than Frank Wild. As an able seaman on Scott's first expedition, he met Shackleton, who invited him to join *his* expedition. Wild accompanied Shackleton to his farthest south. Later, Douglas Mawson, whom Wild met on Shackleton's venture, selected Wild to command the Western Base on his Australasian Antarctic Expedition. As deputy to Shackleton on the Imperial Trans-Antarctic Expedition, Wild was left in charge at Elephant Island. Five years later, in 1921–22, he again sailed with Shackleton, assuming command after Shackleton's death.

LEFT Although one of the greatest Antarctic explorers, Wild never returned there after Shackleton's last expedition. He spent his later years holding a variety of jobs in southern Africa.

Dec. 11ᵉ 1935.

Took our sledging by night today after our experience with soft snow yesterday. Travelled 10 or 11 miles, one both fell over a crevasse. It was weary work and it seemed as though we must be going in the wrong direction for the never ending expanse stretched on forever, yet our sights of yesterday placed us 10 to 15 miles from the head of the Bay of Whales. The night sun cast a weird dull glow over the ice fields, without warmth although out of a cloudless sky, and unlike anything to be found outside the Polar regions. Suddenly I told Kenyon I could see a line of blue water on the horizon. Yes, there it is he said, the Bay of Whales! We had been travelling much too far west.

We made camp. I took off my socks and moccasins, looked at my left big toe and saw it was one big water blister. It had been numb without feeling

ever since our camp up at 6000 ft. I must have either frozen or frost-bitten it there. Well it's a good thing for me that we haven't a trek of 300 miles, our breakfast consisted of fried bacon and boiled dried milk, the first time we have varied our menu.

Dec. 12° 1935

a snare and a delusion! The Bay of Whales? Where? We travelled 12 miles today at 2 miles an hour, but where the water we saw yesterday? No more. The second N. wind on, our trip and the weather misty with no visibility except toward the Ross Sea, over which hung a heavy indigo water sky to the W. of our route of travel. So tomorrow we head in that direction. One little bird dropping in the snow, no other sign of life.

Down to our last quart of fuel now.

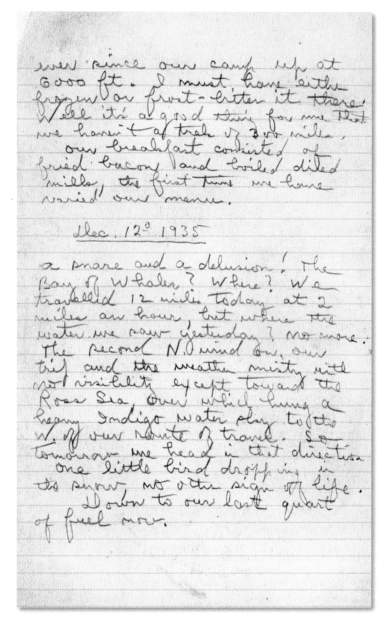

ABOVE Extracts from Lincoln Ellsworth's notebook during the period of his crossing of the Antarctic continent. When Ellsworth's plane finally ran out of petrol, he and Herbert Hollick-Kenyon struggled through terrible weather to walk the rest of the way.

LEFT Lincoln Ellsworth found in both polar regions that long-distant flight meant unexpected delays. In the Arctic, he and Roald Amundsen were forced to land, and could not get airborne again for 24 days.

ABOVE After *Endurance* sank, Shackleton's men attempted to drag the three lifeboats to safety across the ice. It was not an easy task as, when loaded, *James Caird* weighed more than a ton.

BELOW Ernest Shackleton, who was known as "The Boss", in a typical pose. Reaching a point only 97 miles from the South Pole made Shackleton an international hero, earned him a knighthood and put him in a position to organize his Imperial Trans-Antarctic Expedition.

RIGHT Trapped in the ice of the Weddell Sea, *Endurance* was twisted, squeezed, pushed over and eventually crushed. Many of the men hoped she would be released, but Shackleton was realistic enough to admit that: "What the ice gets, the ice keeps."

PAGE 15
CHRISTOPHER COLUMBUS' LETTER
(translation of full letter)

In praise of the most serene Ferdinand, king of Spain, of Bethica (Andalucia) and of the kingdom of Granada, siege, victory and triumph and of the recently discovered islands in the Indian Sea

Ferdinand, King of Spain

Concerning the recently discovered islands in the Indian Sea

Concerning the recently discovered islands

The letter of Christopher Columbus (to whom our state owes much concerning the recently discovered islands in the Indian Sea, for which he was searching diligently for eight months before he had been sent under the auspices of the most invincible Ferdinand King of Spain) sent to the distinguished Lord Raphael Sanchez, Treasury of the same serene king, which that noble and learned man Aliander de Cosco translated from the Spanish idiom into Latin, on the third of the month of May 1493 in the first year of the Pontificate of Alexander VI.

Since the matter of the task undertaken by me has been completed, I wish to express my gratitude to you and I have caused these things to be written down, which remind you about everything of this our journey of sailing and discovery.

On the thirty-third day after departure from Cadiz, I have arrived in the Indian Sea, where I have discovered many islands inhabited by countless men, of which I have taken possession on behalf of our most fortunate king with no one speaking contrary amongst them, to public praise & flags unfurled.

Excelling especially in those things done in the name of the divine Saviour and relying on his help to such a degree, we have arrived at other islands which they call in true Indian Guanahanyn. I have certainly renamed all the rest in a different way with a new name. Indeed another island is now called Sancta Maria Conception, another Fernando, another Isabel, another Juana & in this way I have ordered the remainder to be named. We have landed on this island (which I have said a short time ago should be called Juana). I have gone forth a certain distance from its western shore and thus I have discovered this one to be large with no known boundary, so that I believed the territory to be not an island but the continent Chatay. However, I have seen no towns or municipalities situated near the coast but nearby some villages & rural communities with whose inhabitants I was unable to speak, wherefore as soon as they saw us they stole away.

I was proceeding further, being of the opinion that I was about to discover some town or perhaps houses. At last having certainly progressed far I observed nothing new had come to light. In this way a road was taking us to the North (which I myself desired eagerly to avoid—truly winter reigned in the land) and to journey to the south was more desirable, with less demanding circumstances arising. I have arranged for the others not to be occupied with an advance, but in fact to withdraw.

I have returned: to a certain port (which I had observed): whence I have sent two men of ours inland to search to see if there may be a king in this province or some city. They have walked for three days and have discovered innumerable people and small dwellings but without any ruler, wherefore they have returned. Amongst the things I had already understood from certain Indians, whom I have received in this same place, was how an island of this sort was indeed a province, & thus I have proceeded towards the East, its shores extending further and further to 322 miles, right to the extremities of this island. I have looked towards that other island to the East 54 miles distant from this island of Juana, which I have immediately proclaimed Spanish.

I have withdrawn into this island and arranged the journey to the North, as it were, in Juana island about 564 miles to the East. The island called Juana & other islands in the same region appear very fertile.

I have never seen other comparable great, very safe and wide ports. It is set around with many very lofty mountains & many very large and salubrious rivers flow between them.

Everything on this accessible island is full of very beautifully and distinctly varied species, & has a very large variegated canopy of encircling trees, which I believe never to be free from leaves. So I have indeed seen them becoming green and beautiful as would be usual in the month of May in Spain, wherefore some flowering, some fruitful, some in other conditions following on and without exception were thriving.

A nightingale was singing & other various innumerable sparrows in the month of November when I myself was taking a walk amongst them. There are besides in the said Juana Island seven or eight varieties of palms which, in height and beauty, easily surpass all our trees, plants and fruits. There are both wonderfully fertile fields & extremely desolate plains with different kinds of birds, honey and metals, except iron. In this island, moreover, which we have previously solemnly proclaimed to be Spanish, are large and beautiful mountains, uncultivated areas, farms, forests, and very fertile, peaceful, cultivated fields & most suitable for clusters of buildings.

The advantage of harbours in this island & the excellence of the rivers combined with the health advantages for humans, unless one has seen them, exceed expectations. The pastures and the fruit of these trees differ greatly from those of Juan. Of all other islands I have seen, this henceforth Spanish island abounds in a greater variety of spices, gold and metals & I have knowledge of these things. The inhabitants of both sexes always walk about naked.

PAGE 23
HENRY THE NAVIGATOR'S LETTER
(translation of full letter)

I, Infante Dom Henrique (Prince Henry), Duke of Viseu and Lord of Covilhã, hereby wish to inform those who receive and are shown this letter of release that I issue, and who learn thereof, that I ordered Mr. Heitor de Sousa, overseer of my house and head treasurer, to take charge of all the items he received from me and spent in the said job of treasurer. This was the case every year from the past until now, when this letter is drawn up. Thus, there were gold, silver, money and gold cloth, silk, wool and national and French flax, French linen and tapestries and candlesticks(?), iron, steel, wrought tin and lead and many other things, as well as male and female Moors, parrots and other items that he received and spent when he carried out my orders, relating to my dealings with Guinea. And, altogether, I found that he handled matters well for me, serving me therein in a very good and loyal manner. Nevertheless, I hereby fully release and free him, henceforth forever, as well as all his property, heirs, ancestors and descendents, so that never at any time may he or his heirs, as aforesaid, be required or demanded to give account thereof. It is my wish to grant that, even if by chance at some time it should be found that he owed me money and had an obligation in some of these items that have been mentioned and may be mentioned, that he received in those aforementioned years from that time until now, the same be made irrevocable to him as a reward, which I establish as firmly as I can. I further order and charge my overseers of the estate, accountants and any other members of my staff, never to require him, or likewise any of his heirs, to account for everything he received from me and spent in this way up to this date whereon this letter is drawn up, as has been declared. And for his own protection and that of his heirs I have ordered this release to be given to him. And if anything is lacking therein, because it is not certain how it should be fulfilled, or because of some laws or provisions, whether they are or may be contrary thereto, I have mentioned everything expressly herein, and I beseech His Majesty the King, my Lord, and Infante Dom Fernando (Prince Ferdinand), my dearest and most beloved son, to be so kind as to protect them and ensure that they are protected, and to fulfil this release of mine, in accordance with its contents. And I shall be very grateful to him. Drawn up in my town, on 11th January, by Inácio Baldaia, in the year of Our Lord Jesus Christ fourteen hundred and fifty-eight.

Infante Dom Henrique (Prince Henry)

PAGE 31
CARTIER PATENT
(selected translation)

François, King of France, with God's blessing, and to all those who will see these letters, Hail to thee. Due to the desire to understand and to have knowledge of several countries that one calls uninhabited and owned by primitive people, who live without any knowledge of God and without any need for reason, and at great expense and great risk, we had been sent to discover these countries with several good pilots and well-meaning subjects. To know and experience such countries, to be introduced to the various men who have for a long time been part of our Kingdom, teaching them about love and belief in God and holy law and the Christian doctrine, with the intention of bringing back goodwill to the said countries accompanied by a large number of our subjects, in order more easily to make the other people in these countries believe in our faithful Saint. Amongst others, we sent our dear and well loved Jacques Cartier there, who discovered the great lands of Canada and Ochelago, making a tour of Asia from the west, countries he discovered and from which he brought us back several great riches. The people there are fit and full of spirit and understanding, who, for a long time, we have made live and believe in our faithful Saint with our said subjects. In consideration of this and their good inclination, we decided to send the said Cartier back to the said countries of Canada and Ochelago and to the land of Saguenay if he could reach it, with a good number of ships, and with our aforesaid well-meaning subjects and with the artistic and industrial skills required to explore the said countries further, to talk to the said people, and live with them if necessary in order to achieve our said intention and to do something good for God, our creator and redeemer, spreading his holy and sacred name and Our Blessed catholic church from which we are called. Why is there a need for better order and for sending the said expedition and appointing a "captain general" and master pilot of the said ships, who are responsible for directing these ships and the people, officers and soldiers on them. Let it be known that we were completely trusting of the aforementioned Jacques Cartier, of his senses, self-importance, loyalty, boldness, great diligence and wealth of experience, for whose causes we were driven; appointed and organized by the "Captain General" and master pilot of all the ships and other sea-going vessels for the said venture and expedition for the said state and attending to the "Captain General" and master pilot of these ships and vessels. Having, honouring and practising the honours, prerogatives, precedences, frankness, freedom, security and kindness of the said Jacques Cartier, as long as he liked us, and giving him power and authority to appoint lieutenants, managers, pilots and other agents for the said ships in order to lead them, and in such a number that he had and knew was necessary for the good of the said expedition. And we under the command of the aforesaid, presented what the aforesaid Jacques Cartier took and received to our admiral or vice-admiral, and appointed and requested by the said "captain general" and master pilot. Understanding all his honours, prerogatives, pre-eminences, frankness, liberty, security and kindness, experiencing suffering and enjoyment, and fully and peacefully making use of and obeying him and listening to what he stood for, and all those things affecting and concerning the said state, and taking the little Gallion called *Esmerillon*, which is already old and obsolete, to serve on these ships that needed it and which we wanted the said Cartier to take in order to impress the aforesaid, by which we also mean our provosts in Paris, bailiffs in Rouen, Caen, Orléans, Blois and Tours, Sénéchaux de Maine, Anjou, and Guyenne, and to all our other bailiffs, seneschals, provosts and our other dispensers of justice and officers in both our said Kingdom and our Brittany country, there are no accused prisoners or those accused of any crimes whatsoever in our midst, except for crimes of heresy and divine and human injury to us and deceitful moneychangers that they were unable to give into the hands of the aforesaid Cartier or to his ship's stewards and deputies, for our service in the said venture and expedition, those said prisoners that he knew to be honest, sufficient and capable of serving in this expedition, up to fifty people and based on the choice that the said Cartier would make. Those initially judged and condemned according to their faults and the severity of their wrongdoings; if they are not judged and condemned and satisfaction also previously given to civil and interested parties; for which... however we did not want the release of their people into the hands of the aforesaid Cartier if he found their service slow, but to take the said satisfaction from their goods only. And what release of the said accused prisoners did we want to be made into the said hands of the said Cartier, by our said dispensers of justice and officers ... and with respect to each of them, power and jurisdiction despite opposition or designation to do this, revealed or to be revealed, and without such release in the above manner being in any way different, so that a greater number could not be chosen, aside from the said fifty. We wanted the release of each to be written and certified, in addition to those present and be registered by them, made weak by our loved and loyal chancellor, to know the number and quality of those who would have been bailed and released. Because such is our pleasure. As a witness of what we did, we made our mark with those present. Given to Saint Pris on the seventeenth day of October in the year of grace, one thousand five hundred and forty, and in our twenty-sixth reign. Hereby signed by the King, our lordship the chancellor and others present – Delachesnay and Scellées. Other patent letters are attached to these letters, counter-sealed; the content of which follows. Henri, oldest son of the King, heir apparent of Vienna, duke of Brittany, count of Valentinois and Diois, to our loved and loyal people in our council and chancellery, seneschals, lieutenants and to all our other dispensers of justice and officers in our said Countries and duchy, Hail to thee. We inform you that considering the content of the patent letters from the King, our very honourable lord and father, given in this place of Saint Pris on the seventeenth day of this month, to which these are attached under the counter-seal of our chancellery, you give into the hands of our dear and well-loved Jacques Cartier "captain general" and pilot of all the ships and other sea-going vessels that the King our lord and father...

[handwritten]
sent to Canada and Ochelago and to the land of Saguenay for... [ends]

PAGE 37
TREATY OF TORDESILLAS
(selected translation)

Dom John, by the grace of God, king of Portugal and of the Algarves on this side and beyond the sea in Africa, lord of Guinea with Ruy de Sousa, lord of Sagres and Berenguel, Dom Joao de Sousa, his son, chief inspector of weights and measures of the said Most Serene King our brother, and Ayres de Almada, magistrate of the civil cases in his court and member of his desembargo it was treated, adjusted, and agreed for us and in our name and by virtue of our power with Don Ferdinand and Dona Isabella, by the grace of God king and queen of Castile, Leon, Aragon, Sicily, Granada, Toledo, Valencia, Galiciaj Majorca Seville, Sardinia, Cordova, Corsica, Murcia, Jaen, Algarve, Algeciras, Gibraltar, and the Canary Islands, count and countess of Barcelona, lord and lady of Biscay and Molina, duke and duchess of Athens and Neopatras, count and countess of Roussillon and Cerdagne, marquis and marchioness of Oristano and Gociano, together with the Prince Don John, their very dear and very beloved first-born son, heir of their aforesaid kingdoms and lordships in regard to the controversy over what part belongs to us and what part to the said Most Serene King and Queen our brother and sister, of that which up to this seventh day of the present month of June, the date of this instrument, is discovered in the ocean sea, in which said agreement our aforesaid representatives promised among other things that within a certain term specified in it we should sanction, confirm, swear to, ratify, and approve the above-mentioned agreement in person: we, wishing to fulfill and fulfilling all that which was thus adjusted, agreed upon, and authorized in our name in regard to the above-mentioned, ordered the said instrument of the aforesaid agreement and treaty to be brought before us that we might see and examine it, the tenor of which, word for word, is as follows:

In the name of God Almighty, Father, Son, and Holy Ghost, three truly separate and distinct persons and only one divine essence. Be it manifest and known to all who shall see this public instrument, that at the village of Tordesillas, on the seventh day of the month of June, in the year of the nativity of our Lord Jesus Christ 1494, in the presence of us, the secretaries, clerks, and notaries public subscribed below, there being present the honorable Don Enrique Enriques, chief steward of the very exalted and very mighty princes, the lord and lady Don Ferdinand and Dona Isabella, by the grace of God king and queen of Castile, Leon, Aragon, Sicily, Granada, etc., Don Gutierre de Cardenas, chief auditor of the said lords, the king and queen, and Doctor Rodrigo Maldonado, all members of the council of the said lords, the king and queen of Castile, Leon, Aragon, Sicily, Granada, etc., their qualified representatives of the one part, and the honorable Ruy de Sousa, lord of Sagres and Berenguel, Dom Juan de Sousa, his son, chief inspector of weights and measures of the very exalted and very excellent lord Dom John, by the grace of God king of Portugal and of the Algarves on this side and beyond the sea in Africa, lord of Guinea, and Ayres de Almada, magistrate of civil cases in his court and member of his desembargo, all of the council of the said lord King of Portugal, and his qualified ambassadors and representatives, as was proved by both the said parties by means of the letters of authorization and procurations from the said lords their constituents, the tenor of which, word for word, is as follows:

[Here follow the full powers granted by Ferdinand and Isabella to Don Enrique Enriques, Don Gutierre de Cardenas, and Dr. Rodrigo Maldonado on 5 June 1494; and the full powers granted by John II. to Ruy de Sousa, Joao de Sousa, and Ayres Almada on March 8, 1494.]

"Thereupon it was declared by the above-mentioned representatives of the aforesaid King and Queen of Castile, Leon, Aragon, Sicily, Granada, etc., and of the aforesaid King of Portugal and the Algarves, etc.:

[I.] That, whereas a certain controversy exists between the said lords, their constituents, as to what lands, of all those discovered in the ocean sea up to the present day, the date of this treaty, pertain to each one of the said parts respectively; therefore, for the sake of peace and concord, and for the preservation of the relationship and love of the said King of Portugal for the said King and Queen of Castile, Aragon, etc., it being the pleasure of their Highnesses, they, their said representatives, acting in their name and by virtue of their powers herein described, covenanted and agreed that a boundary or straight line be determined and drawn north and south, from pole to pole, on the said ocean sea, from the Arctic to the Antarctic pole. This boundary or line shall be drawn straight, as aforesaid, at a distance of three hundred and seventy leagues west of the Cape Verde Islands, being calculated by degrees, or by any other manner as may be considered the best and readiest, provided the distance shall be no greater than abovesaid ...

DOCUMENT PAGE 70
VITUS BERING REPORT
(translation of full letter)
To the brilliant count
My merciful patron
Nikolay Fedorovich

Although your Majesty should know through my previous reports the reason, why the expedition, that was entrusted to me, was not moving forwards for such a long time, but even now I cannot bear to inform your Majesty, that despite all the efforts and means I had tried to continue the expedition without any prolongation, there was no opportunity to go to sea earlier. It happened because from the beginning of the expedition until now we lacked provisions. We also had difficulties with food-boats building and with provisions' transportation both by sea and by land, all that stood in the way of doing our best to continue the expedition. Besides that all these shortages had not been removed before the expedition started, and they started to improve only after having come to Siberia with (here I need to mention) so big a staff: apart from two trips to Yakutsk there were 300 people. And also, besides those who had been sent from St Petersburg, about 230 Siberian carpenters and smiths were equipped for a huge ship-building. Thus, it was impossible to bring easily enough food for so many people by these bad roads. We also had another difficulty that we managed to overcome only by now: to fit out the last two packet-boats with two el-boats and two launches to them. And on packer-boats, on double-launch, taken by Captain Shpanberkh's crew that will return here again and on a boat borrowed from the governing hunt we have shipped food for one year and eight months that we hardly managed to bring here. And on these boats with God's help I will go from here to Kamchatka, to Avachinskaja guba as soon as I can. There I will stay for winter. Next year, 1741, I will set out on the trip that is in my instructions as soon as possible. But I don't see, I am not sure and I can't assure your Majesty that both my and captain Shpanberkh's trips will be effective and could be made without troubles, caused by the lack of food and its hard transportation, the lack of important materials that can not be found here and the other lacks, as this land is desert and doesn't contain everything we need. And having informed your Majesty about it I stay your Majesty's obedient servant

V Bering
29 August 1740

PAGE 82
CHINESE PASSPORT OF NEY ELIAS
This passport, issued by Mr. Wade, British Envoy to China, is for four of the envoy's subordinate officials with attaché to enter the country by way of the Yunnan Province from Burma, and then return by the same way or travel to Shanghai along the Yangtze River, either route to be decided at the appropriate time, and also for governors of the Yunnan and Guizhou Provinces and of the other relevant provinces to notify their subordinate customs officers and local officials to welcome them and help them with their journey. The names of the officials in question are listed as follows, along with the Ministry of Foreign Affairs' stamp.

The officials are:
Bailewen, 3rd rank military officer
Yilaiyashi, 4th rank civil official
Anderson, 4th rank medical official

Date: 31 July, 1874
18th day, 6th month, 13th year of Emperor Tongzhi's reign

PAGE 181
DUMONT D'URVILLE REPORT
(translation of full report)
Sent by Commodore d'Urville   Rec from
Sir J Franklin
                                         June 10
To Captain Parker A.D.C. for insertion in
the Hobarton newspaper 20th February 1840

Expedition
of the French Corvettes Astrolabe and Zélée

under the command of Commodore d'Urville and Captain Jacquinot.

Every inhabitant of Hobart-Town could remember that the two corvettes Astrolabe and Zélée, under the command of Captain Dumont d'Urville, which have just returned to our port, had already spent the last fifteen days of December there, and had even left behind several sick people in this town, who they had to pick up on their return.

After long and hard journeys, completed with unbelievable success, in the Antarctic region to the South of Cape Horn, through most of the archipelagos of Oceania and through the whole of Malaysia, these two ships had been attacked by dysentery during the passage from Sumatra to Van Diemen's Land. Several people succumbed to it and several others were still extremely ill.

Despite this setback, Commodore d'Urville wanted to undertake a second expedition to the polar region, to identify where the ice stops in this part of the world, and particularly to collect definitive data on the true position of the Magnetic South Pole.

Today, he is back from this journey, after being more successful than he could have wished for. In the interests of both humanity and science, and to prevent any erroneous assumptions and absurd exaggerations, Commodore d'Urville wanted us to have the following note, which is a correct and concise account of the main achievements.

The expedition left Hobart-Town on 2nd January, and aided by constant winds from the W.S.W. to the W.N.W, was able to head towards an area 45 miles to the S/SW.

On 15 January, at 59° lat. South, Cook's route was crossed, the only known navigator to this day to have gone so far in these parts. The next day, at 60°, the first ice floes were encountered. From then until the 19th, their number and size gradually increased, becoming almost innumerable. That evening, at 66°, land from the South to W.S.W. could be seen as far as the eye could see on either side.

After having passed very close to huge chains of enormous icebergs, on the 21st, the two corvettes could see land five or six nautical miles away and carried on in superb weather, with a light breeze from the East. The same day, in the evening, all magnetic observations were successfully completed on a huge slab of ice, and two small craft went ashore to collect several samples of rock from a point on the coast left uncovered by the ice.

Commodore d'Urville wanted to continue his exploration of these parts, but from 23 January, the passage was blocked by an ice field stretching from the land to the North for an unknown distance. He had to tack to get out of the ice in which he was caught up. The next day, the expedition had to endure a harsh East gale, which posed the greatest danger to both Corvettes, but particularly Zélée, the loss of which was threatened for a long time by the large icebergs that bordered the ice field.

However, the ships, after tremendous effort and extreme tiredness, managed to get out of this dangerous strait. Commodore d'Urville continued his journey to the West, trying to get close to land, as far as it was possible. But nothing could be seen apart from the ice field, and on the 30th, an area of approximately twenty miles with a steep, solid and compact slope, approximately 150 feet high, could be seen very close. However, land itself could not be seen anywhere. Moreover, the horizon was still so hazy that it was impossible to make anything out from a distance of more than ten or twelve nautical miles. Further on, the ice field that stopped any subsequent progress towards the South was encountered again.

On 1 February at 65° 20' lat. South and 131° long. East (Greenwich)", the meridian was passed without declination, the variation had become N.W. from N.E., which it had been a few days previously, and magnetic observations were made both on land and at sea, which allowed the location of the magnetic pole to be established with the absolute accuracy desired. Consequently, Commodore d'Urville, deeming the task they had been set to be fulfilled, continued the journey from Hobart-Town where he returned on 17th February in the evening, after 46 days away.

Despite the cold, tiredness and the dangers endured, the two crews returned in good health, apart from a few sailors on the Zélée who had been sick since departure.

Commodore d'Urville named the newly discovered land Adélie Land. The area identified, an area covering approximately 150 nautical miles, is between 66 and 67 degrees Lat. South on one side, and between 136 and 142 degrees long. E. on the other. The average height was 130 feet above the horizon. The snow and ice that covered it almost levelled its surface, only leaving ravines, bays and headlands. No sign of any vegetation. Around the coasts, a few large whales and dolphins, a few sealskins, very few penguins and a few petrels and albatross could be seen. That was everything the animal kingdom had to offer. Captain d'Urville thought the area was much larger. The ice fields prevented any progress towards the West as things stood then, but to the East, or rather the S.E., it did not seem impossible to go further, insofar as the eye could see from the top of the masts. The unfavourable and cold East winds, and the feeling that he had enough data to establish the Magnetic Pole, were what prevented him from trying a longer journey to the E.S.E. He really hopes that other navigators will be able to take the explorations that had already begun even further.

Further to this note, Commodore d'Urville informed us that he wanted us to publicly express his gratitude for the courtesy and generous hospitality which his journey companions and himself had received from Governor Sir John Franklin, from Lady Franklin, and all the civil and military authorities in the Colony. He said that he experienced complete kindness everywhere, and an honourable willingness to satisfy his wishes for the success of the journey he was directing. Finally, he wanted to express his real and sincere affinity to the research that he was endeavouring to accomplish in the interest of science and navigation.

For him, he has just experienced again the kindly welcome he had received from Hobart-Town, approximately twelve years ago, and he will bring back this lovely memory to his homeland.

PAGE 192
ROALD AMUNDSEN LETTER
The Fram Expedition
15 Dec 1911

Your Majesty.

I herewith allow myself to announce, that 5 men of the Fram Expedition – myself included – arrived here to the South Pole area – according to obs. 89° 57' 30' S. Lat. – yesterday Dec. 14th – after a successful sled ride from our winter station "Framheim". We left this on the 20th of Oct. with 4 sleds, 52 dogs and viands for 4 months. We have on our way decided the expanse of the great "Ross Barriers" to the south – ca 86° S. Lat. – as well as the joining of King Edward's Land and Victoria Land in the same place. Victoria Land ceases here, whilst King Edward's Land continues in a SWerly direction until ca 87° S. Lat. with a mighty mountain chain with peaks up to 22,000 f.a.w. These conjuncting mountain chains I have allowed myself to name – as I hope with approval – "Queen Maud's chains". – The great inland plateau we discovered – at ca 88° S. Lat. – going over into a fully flat highland plain, which again at 89° S. Lat. fairly gently began inclining down towards the other side. The height of the plain here is ca 10,750 f. We have today with a radius of 8 km circled the geographical south pole, raised the Norwegian flag and named the gently inclining plain, upon which we have successfully decided the location of the geographical south pole, "King Haakon VII's Plain" with – as I hope – Your Majesty's approval.

Tomorrow we begin the journey back with 2 sleds, 16 dogs and well equipped with viands.

In deference,

Roald Amundsen

INDEX

CREDITS

The publishers would like to thank the following sources for their kind permission to reproduce the pictures in this book. Key: t=top, b=bottom, c=centre, l=left and r=right.

AKG-Images: 40, 64tl, 84, 86t, 89b

Alamy: 79

Archives Municpales de SaintMalo: 31l

Art Gallery of South Australia, Adelaide: 141c

Yann Arthus-Bertrand: 103t

Bridgeman Images: 12, 24b, 25t, 26, 30tr, 30cr, 30br, 33b, 36b, 44tl, 44tr, 46b, 47bl, 51t, 53, 62b, 63b, 73bc, 78tl, 81t, 96t, 102br, 103b, 133t, 179tl, 179cl

Getty Images: 8t, 21t, 43tl, 43tr, 46t, 48br, 64tr, 70b, 105, 106b, 134t, 134b, 140t, 145cr, 145br, 151, 152c, 170, 172bl, 174b, 175tl, 191t

Lane Kennedy: 11t

Library of Congress: 171cl, 202b

Lonely Planet Images: 35b, 62t, 89t, 128b, 146r, 179br

Mary Evans Picture Library: 13, 15b, 19br, 20b. 22l, 30, 37b, 47tr, 68b, 73bl, 74tl, 97c, 104bl, 104br, 106t, 111b, 116t, 116b, 119t, 120b, 128tr, 133bl, 133br, 147b, 159, 173, 181b, 184tl

Mawson Centre, South Australian Museum: 194bl, 195tl, 197b

Missouri Historical Society: 47b

National Archives & Records Administration: 47tl, 50b

National Gallery Art of Washington: 33t

National Geographic Image Collection/George F.Mobley: 111t

National Library of Australia: 142c, 142b

National Maritime Museum: 9, 70t, 130l

National Parks Service Museum Management Program & Valley Forge NHS: 46c

National Portrait Gallery, London: 128tl, 155bl, 155bc, 160c

Naval Historical Foundation: 168b, 171tl

REX/Shutterstock/Picture-Desk/The Art Archive: 8-9, 20t, 22b, 27t, 34t, 52, 95, 97b, 141c, 145cl, 184tr

Royal Geographical Society: 6, 11b, 14b, 15t, 16-17, 19t, 19bl, 21b, 22l, 27b, 28-29, 43b, 48t, 51b, 54t, 55t, 56t, 56b, 57, 60-61, 63t, 66-67, 72, 73br, 74tr, 74b 75, 76b, 77, 78tr, 80t, 81c, 81b, 82, 82-83, 85, 86b, 87, 90b, 91, 94, 98, 99, 100, 101, 102b, 107t, 108, 109, 110r, 112, 113, 114, 115, 116c, 118l, 118r, 119b, 120tl, 120tr, 122r, 123c, 123b, 124-125, 130r, 136b, 137, 139t, 139b, 140t, 144, 150, 152t, 155r, 156, 157, 158, 160t, 161, 163, 164, 165, 166b, 167, 168t, 169t, 170t, 172t, 172br, 175tr, 175b, 180, 181t, 182, 183, 184b, 185b, 186-187, 188, 189, 190, 197c, 197cr, 198, 199, 200r, 201, 202t, 203

Staattsbibliothek zu Berlin: 55b

State Library of New South Wales: 194tl, 194br, 195, 197t

Topfoto: 64b, 78b, 107b, 131b, 135b, 136t, 147l, 147r, 152b, 163t, 163b, 171b

Ullstein Bild: 199

Mark Walker: 10, 14t, 18, 24t, 32, 36t, 42, 44b, 50t, 54t, 65, 68t, 76t, 80b, 88, 90t, 97t, 102t, 110l, 117, 122tl, 131t, 135t, 138t, 143, 146b, 153, 154, 162, 166t, 171tr, 174t, 179, 184t, 191b, 192, 193, 196, 200l

Every effort has been made to acknowledge correctly and contact the source and/or copyright holder of each picture and Carlton Books Limited apologises for any unintentional errors or omissions which will be corrected in future editions of this book.